MATHEMATICS

Marvels and Milestones

(Queries and Answers)

Preface

1. What is the object of writing this book ?

To transport the reader to the highest level of mathematical awareness and to acquaint him with outstanding mathematical achievements.

2. Which mathematical achievements are being alluded to ?

The foremost of them all is the pluralization of mathematics, i.e., where we had geometry, we now have geometries, and algebras rather than algebra, and number systems rather than number system.

Some others are:

Galois' theory of algebraic equations;

Godel's incompleteness theorem;

Fourier's series and infinite sets;

Group Theory; Matrices; complex analysis;

Topology; Functional analysis;

etc.

3. For whom is the book intended ?

It is intended for the intelligent layman who is in search of short and pointed answers to his queries but is little inclined to undertake detailed study of mathematical ideas and concepts.

4. Does it hold any appeal for young students ?

Yes. Here they can have a glimpse of mathematics beyond what they had studied at the school.

5. Is the book intended to be a substitute for some text book ?

No, not at all. The aim is far too modest. It is to spur the reader to go on to fuller accounts than given here.

6. Is the book of any interest to the mathematician ?

A mathematician is usually confined to his special but limited field of interest. This book will provide him with an overall view of mathematics.

This book will also assist him in seeking answers to his philosophical uncertainties in mathematics. Incidentally every discipline has such uncertainties.

7. The book has three broad divisions. Should they be read in the sequence in which they are given in the book ?

No, not necessarily. Any order may be adopted.

The questions also need not be read consecutively unless they interest the reader. If something seems uninteresting or unattractive at first glance it may be skipped.

One could come back to it if one thought it was still interesting.

8. Why has the question answer form of description been adopted?

It is because long narratives soon tire down the patience of the general reader whereas question answer form helps to sustain his interest.

9. What pattern is followed in the sequence of questions ?

As far as possible logical pattern is followed, by which is meant that a question is either suggested or anticipated by the previous question or it may be a related question.

10. What is the style of presentation ?

The answers are in simple, lucid and easy to understand language, and meant to be brief as far as possible.

11. But what if sometimes detailed answers are unavoidable ?

In such cases the answers have been split into small paragraphs after one or more of which one could skip the rest according to taste and patience.

12. What is the basic requisite for reading the book ?

Love for mathematics and for things mathematical.

13. What should be the mathematical background for reading this book ?

Not much. Knowledge of elementary mathematics is enough.

14. What are the main topics in the chapter on Geometry ?

It includes the following :

(i) Euclidean geometry and allied concepts.
(ii) Lobachewskian and Riemannian geometry.
(iii) Geometry of the Earth, space and Elementary particles.
(iv) Projective Geometry.
(v) Coordinate geometry of 2, 3, 4 and n dimensions.
(vi) Geometry of colour space.
(vii) Finite geometry.
(viii) Topology.
(ix) Problem of the Bridges of Koenigsberg.
(x) Four colour problem.
(xi) The axiomatic method in geometry.
(xii) Hilbert's Formalism.
(xiii) Godel's discovery.

15. What are the main topics in the chapter on Algebra ?

It includes the following:

(i) Arithmetic as abstraction.
(ii) Arithmetic as the theory of numbers.
(iii) Extension of the number system.
(iv) Algebra as the Theory of Equations.
(v) Galois' Theory of Equations.
(vi) Diophantine Equations.
(vii) Abstract Algebra.
(viii) Theory of Groups and related matters.
(ix) Rings, Vectors, Matrix, Integral Domain, Field, Vector space, Linear Algebra.
(x) Hilbert space, Banach space.
(xi) Boolean Algebra.
(xii) Russell's epigram of 1901.
(xiii) Countable and Uncountable sets.
(xiv) Continuum Hypothesis.
(xv) Barber's Paradox.
(xvi) Russell's Paradox.

17. How is the book to be read?

First read the question. If you think you do not know the answer, just go ahead. But if you think you have an idea of the probable answer, then pause for a moment to guess what it could be. Next read the answer and see whether your guess was correct. If you were right, it means a small joy and relief, but if not, you have the correct answer at hand.

18. What are the sources of the contents of this book?

In a book like this it is not possible to recall where was a particular idea or concept first encountered. The author is therefore indebted for ideas expressed herein to more people than he can enumerate.

19. Any thanking or acknowledgements?

The publication of this book has been possible on account of the continued interest by my daughter Mrs. Kiran Bhatt, and the persistent and untiring efforts of Dr. Mrs. Latika Jha. They deserve special thanks and appreciation. But I do not want to thank them. I dedicate to both of them this First Edition of the book.

My sincere thanks go to the editorial and production teams of Messrs. ABD Publishers, especially Shri Gopal for keen insight and unfailing co-operation during the design and production of the book.

20. How about suggestions for improvement of the book?

Comments and suggestions for improvement of the book are welcome.

A. L. Audichya

Contents

1

Geometry and Geometries

1. **How many geometries are there ?**

Mainly three. But there could be several.

2. **Which three are they ?**

Euclidean geometry, Lobachewskian geometry and Riemannian geometry.

3. **Does anything peculiar make them different from one another ?**

Yes. In Euclidean geometry, the sum of the measures of the three angles of a triangle is always equal to 180°, but in Lobachewskian geometry, it is always less than 180°, whereas in the case of Reimannian goemetry it is always greater than 180°.

4. **Then the three must be in constant conflict with one another !**

No, they co-exist in a relatively peaceful setting.

5. **What is Euclidean Geometry?**

The geometry one reads at school in which figures or diagrams are drawn on a peice of paper or on an ordinary black-board is called Euclidean geometry in honour of the mathematician Euclid.

He lived about 300 B.C. in Syria but was of Greek origin.

6. **What did Euclid do for Goemetry ?**

He put together all the material of geometry accumulated up to his time in a systematic and logical form and compiled it in thirteen books called 'Elements'.

He developed geometry as a logical structure.

7. **What is a logical structure ?**

In a logical structure a few undefined terms and a few unproved propositions are assumed, and all the rest in developed by logic.

The undefined terms are called basic concepts, and the unproved propositions called "self-evident truths", axioms, postulates or simply assumptions.

8. How can undefined terms and unproved propositions find place in any logical structure ?

In any systematic study it is natural to expect that we carefully define all our terms so that we know what we are talking about. But since every term has to be defined by somthing which has already been previously defined, and this must itself be defined, and so on; this backward journey must stop somewhere. Hence there are a few undefined terms which are taken as obvious and for which definitions are not required.

Likewise, to prove that a theorem is true is to show that it follows from some previously proved propositions, and these in turn must themselves be proved, and so on. This backward journey must again stop somewhere so that there are some propositions which are accepted as true and for which proof is not required.

9. Are the unproved propositions or postulates subject to no restrictions at all ?

They are subject to two important restrictions. The first is that the postulates must be *consistent*. This means that contradictory statements are not implied by the postulates. They must not lead to 'A is B' and 'A is not B'.

The second is that the postulates must be *complete*. This means that every theorem of the logical system must be deducible from the postulates.

10. Any other constraints ?

It is preferable that the postulates be *independent*. It means that no postulate should be logically deducible from the others.

This is desirable for reasons of economy and elegance but inclusion of a not independent postulate does not invalidate the system. Detection of such a postulate is sometimes not quite easy, either.

And, of course, the postulates must be simple and not too many in number, otherwise the logical system developed will be of little gain.

11. Should not the postulates be in aggrement with everyday experience ?

The postulates need not necessarily be in agreement with everyday experience, because development of a structure on the foundation of new and bold assumptions only can lead to fresh discoveries and important advances.

That extremely bold assumptions led to the discovery of geometries other than Euclidean is a case in point as we shall presently see.

12. How are postulates used and to what end ?

The few assumptions or rules made in the beginning are so ordinary and inevitable that it is impossible to foresee their consequences. From these rules is drawn a consequence which must be granted, and from this consequence another, and so on till the final result is reached, which is often surprising.

One feels a strong impulse to re-examine the sequence of ideas but that only re-affirms the final result !

13. What are the basic concepts of Euclidean Geometry ?

In Euclidean geometry the point and the line are the basic concepts. A point is said to have no magnitude, and a line has no breadth.

But these are suggestive descriptions and not mathematical definitions.

14. How are points and lines of geometry different from their physical counterparts ?

The concept of a point is that of a very small object whose physical manifestation is a pencil dot. A straight line manifests itself in a stretched thread or a ray of light.

Points and lines of geometry are abstractions from the pencil dots and pencil lines of everyday experience.

15. What is the use of such abstractions ?

The advantage gained from such abstractions is that points and lines of geometry have far simpler properties than physical dots and lines. For example, two big enough pencil dots can be joined by many pencil lines, but if the dots become smaller and smaller in size, all the lines look almost alike and we have no difficulty in apreciating the axiom of geometry that one and only one straight line can be drawn through any two points.

16. What are the postulates of Euclidean Geometry ?

Euclid's postulates are as follows :

1. A straight line can be drawn from any point to any other point.
2. A finite straight line can be extended continuously in a straight line.

3. Given any point and any distance, a circle can be drawn with that point as its centre and that distance as its radius.
4. All right angles are equal to one another.
5. If a straight line crosses two other straight lines so that the sum of the two interior angles on the one side is less than two right angles, then the two other straight lines, if extended far enough, cross on that same side of the first line where those angles are.

17. What are the axioms of Euclidean geometry ?

Euclid's axioms are as follows :

1. Things which are equal to the same thing are also equal to one another.
2. If equals be added to equals, the wholes are equal.
3. If equals be subtracted from equals, the remainders are equal.
4. Things which coincide with one another are equal to one another.
5. The whole is greater than the part.

18. How are axioms different from postulates ?

Euclid's distinction between axioms and postulates is not retained by modern writers, who use these terms interchangeably and call them basic assumptions.

19. What did Euclid achieve from such a small set of basic assumptions ?

Using these few basic assumptions only, Euclid proved hundreds of theorems, many of them profound, and brought about the ordering of theorems.

The idea of proof itself, which constitutes the very spirit of mathematics, was introduced by Euclid.

Since proofs must all be worked out within the framework of the assumptions, Euclid's choice of the basic assumptions is indeed remarkable and undoubtedly the work of a genius.

20. What is the parallel postulate ?

Euclid's fifth postulate mentioned above is known as the parallel postulate. An equivalent form of the postulate is the following :

"Through a given point, which is not on a given line, one and only ***one*** *line can be drawn parallel to the given line."*

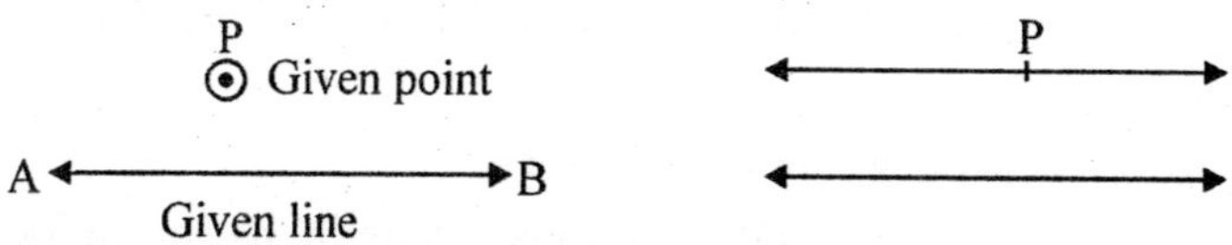

This is known as the famous "parallel postulate". It is a mark of Euclid's genius that he recognised its necessity.

A logical consequence of this postulate is the Theorem of Pythagoras which says that the sum of the angles of a triangle is equal to two right angles.

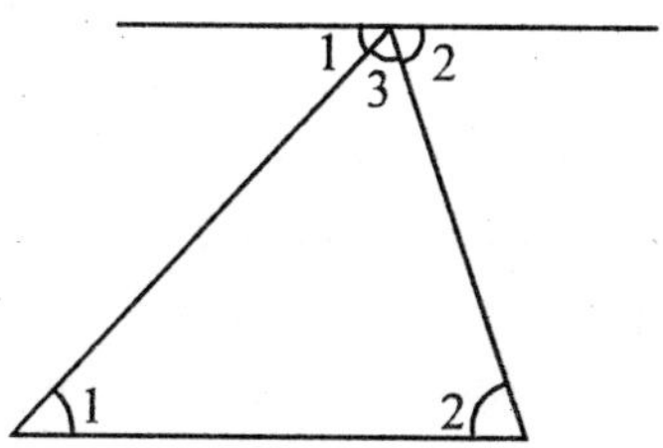

$\angle 1 + \angle 2 + \angle 3 = 180°$

21. What is Labachewskian Geometry ?

The postulate mentioned above appears to be so obvious that one never feels it can or may be changed. But a few mathematicians, lobachewsky being one of them, thought of seeing what happened if the postulate were replaced by the following one :

" Through a given point which is not on a given straight line, ***two*** *different lines can be drawn both of which are parallel to the given line".*

We could try to draw a figure like the following, where two distinct lines are drawn through the point P, one to the right and the other to the left.

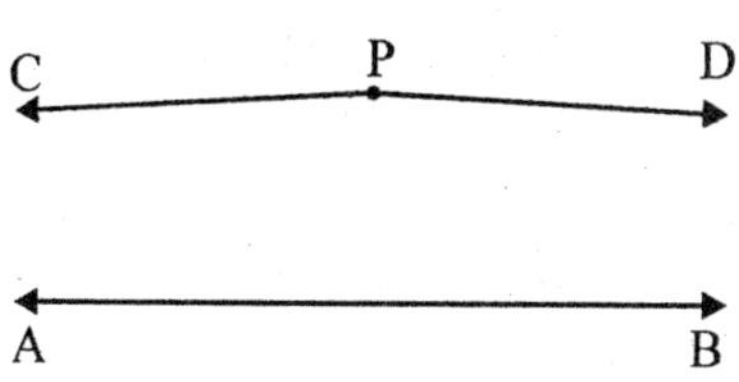

Mathematicians found that not only no fallacy resulted from this strange assumption but a logical consequence of the new assumption led them to a new geometry in which the sum of the angles of a triangle is *less* than 180°.

22. Is it not a strange assumption ?

Logically speaking, there is nothing wrong in the assumption as one is free to select any basic assumptions so far as they do not contradict each other.

23. But the two lines in the figure do not appear parallel to the given line !

The reason why the two lines in the figure above, one to the right and other to the left, do not appear parallel to the given line is that the figure has beeen drawn in an ordinary plane, where only Euclidean geometry holds good and not the new geometry !

24. Who else hit upon the new idea ?

Three different mathematicians, Gauss of Germany, Bolyai of Hungary and Lobachewsky of Russia discovered this logically consistent geometry quite independently and at about the same time, about 1826.

25. Why then is the geometry called Lobachewskian ?

Gauss the most famous mathematician of that time did not venture to come out with these ideas into the open for fear of damaging his reputation.

Bolyai had complete courage of his conviction, but he did not develop the new ideas as deeply and fully as Lobachewsky.

Lobachewsky was the first to express the ideas openly, and also developed them subsequently in a number of papers. The new geometry is, threfore, called Lobachewskian geometry.

26. What is Riemannian Geometry ?

Riemann, a German mathematician, at about 1854, thought of replacing the parallel postulate by the following one :

"Through a given point which is not on a given straight line, ***no*** *line can be drawn which is parallel to the given line."*

A logical consequence of this assumption led him to a geometry in which the sum of the angles of a triangle is *greater* than 180°.

This geometry is known as Riemannian Geometry.

27. Which theorems are true in all the three geometries ?

Theorems of Euclidean geometry which do not depend upon the parallel postulate remain unchanged. For example, the following theorems hold good in all the three geometries :

(i) Vertically opposite angles are equal.

(ii) Base angles in an isosceles triangle are equal.

28. What are the differences ?

The following comparison brings out the differences clearly.

In Euclidean geometry :

(i) The sum of the three angles of a triangle is always *equal* to 180°.

(ii) Parallel lines never meet, no matter how far extended, and always remain a constant distance apart.

(iii) Two triangles may have the same angles but different areas. Such triangles are called similar triangles, and one is the magnification of the other.

(iv) Through a point outside a straight line, only one perpendicular can be drawn to the straight line.

(v) The ratio of the circumference of a circle to its diameter is equal to π.

In Lobachewskian geometry :

(i) The sum of the three angles of a triangle is always *less* than 180°, and the amount by which it is less is proportional to the area of the triangle.

(ii) Parallel lines never meet, but the distance between them becomes less as they are further extended.

(iii) Only triangles equal in area can have the same angles so that triangles having unequal areas can never be similar. In this geometry, as a triangle increases in area, the sum of its angles decreases.

(iv) Through a point outside a straight line, only one perpendicular can be drawn to the straight line as in the Euclidean geometry.

(v) The ratio of the circumference of a circle to its diameter is always greater than π, and the ratio is larger the larger is the area of the circle.

In Riemannian geometry :

(i) The sum of the angles of a triangle is always *greater* than 180°.

(ii) Every pair of lines in the plane must intersect.

(iii) The angles increase as the triangle grows larger.

(iv) Any number of perpendiculars can be drawn from an appropriate point to a given straight line.

(v) The ratio of the circumference of a circle to its diameter is always less than π and decreases as the area of the circle increases.

29. Which is the true geometry ?

Each geometry is true but only on the surface for which it is meant.

Euclidean geometry applies to figures drawn on a piece of paper or on a flat surface.

The non-Euclidean geometry of Riemann is very nearly true for figures drawn on the surface of a sphere.

The non-Euclidean geometry of Lobachewsky holds good for figures drawn on a surface called pseudo-sphere. See below :

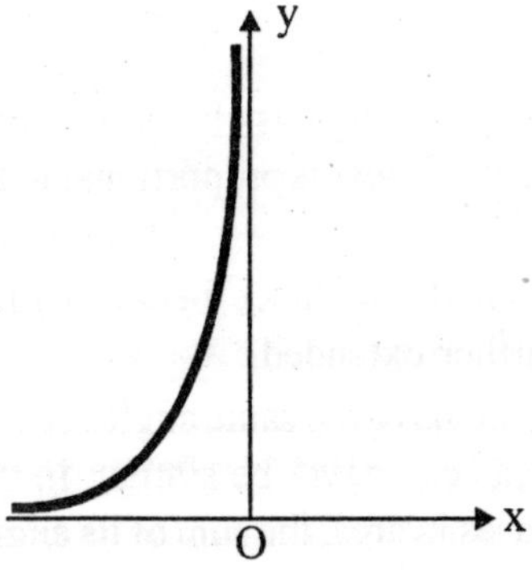

Tractrix
The thick line denotes the curve

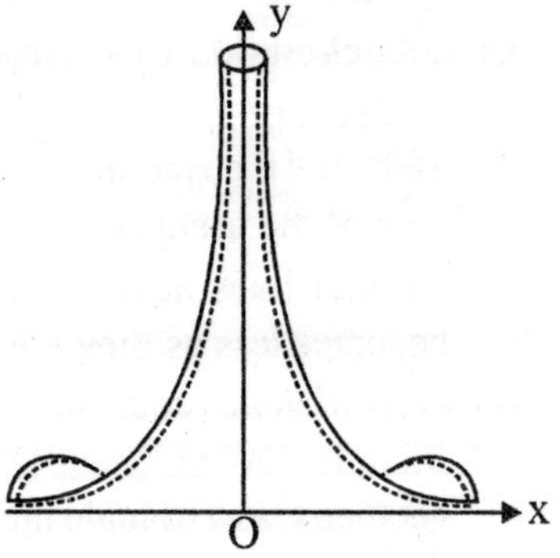

Pseudo-sphere
Generated by revolving the curve tractrix around Oy

A pseudo-sphere is the surface of revolution obtained by revolving the curve known as the tractrix about a vertical line Oy.

Triangles drawn on different surfaces are shown below :

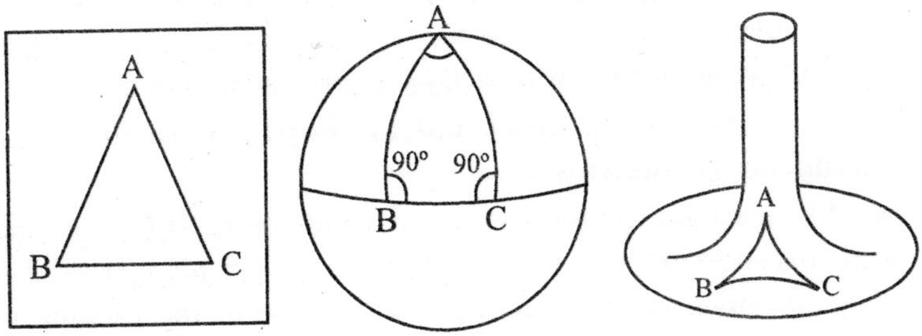

Flat surface	Sphere	Pseudo-sphere
∠A + ∠B + ∠C	∠A + ∠B + ∠C	∠A + ∠B + ∠C
Equal to 180°	Greater than 180°	Less than 180°

Different geometries hold good on different surfaces.

30. Since a geometry is created solely on its axiom system, what is its dependability in the physical world ?

The properties of our physical space are so accurately defined by Euclidean geometry in its routine applications that for more than two thousand years it continued to be identified with absolute truth about the physical space.

Only with the discovery of non-Euclidean geometries came the realization that geometry is not the truth about physical space. It is only a study of posible spaces.

Different geometries, determined by different axiom systems are, therefore, not descriptions of reality.

They are mere models.

From this point of view it is sheer luck that the Euclidean model describes physical reality in quite a large setting.

31. What then is meant by the truth of a mathematical theorem ?

A mathematical theorem is essentially a conditional assertion.

It is true only if the set of postulates from which it is derived is true.

But whether the set of postulates itself is true or false is not asserted by the theorem.

32. Why ? What is the reason ?

The reason is that the postulates are formulated in terms of concepts, which strictly speaking, have no specific meaning, so that whether the postulates are true or false cannot be asserted.

33. Is the Euclidean geomtry not in conflict with the non-Euclidean geometries ?

No. Since a plane has zero curvature, if zero be substituted as the value of curvature in the formulae of non-Euclidean geometries, the resulting formulae become identical with those of the Euclidean geometry.

Euclidean geometry can, therefore, be viewed as a particular case of the non-Euclidean geometries, which are more general.

34. What is meant by a straight line ?

One thing that immediately comes to mind is that straight lines on the surface of a sphere or on that of a pseudo-sphere are really curved and it doesn't appear proper to call them straight.

But it all depends upon how we define a straight line.

One way of defining a straight line is to identify it with the *shortest distance* between two points.

35. How does this simplify the matter ?

Now the shortest distance between two points *on* the surface of a spehre is not a staight line but is a portion of the arc of a circle lying on the surface of the sphere.

Such a circle is called a "great circle" and has its centre at the centre of the sphere*.

Generalising this notion, the curve which lies on a surface and is the shortest distance between two points on the surface is called a 'geodesic' of that surface.

What straight lines are for planes, geodesics are for surfaces.

---

* It the two points on the surface of the sphere are joined with the help of a ruler piercing through the sphere as it were, the line so obtained will no longer be *on* the surface of the sphere.

But since the line must be on the surface, it has to be along the "great circle".

A great circle cuts the sphere in two equal parts. The Equator is a great circle, but the circles of latitudes are not. A longitude is half of a great circle.

36. Do geodesics differ from surface to surface ?

Yes, geodesics differ from surface to surface.

Geodesics on a plane are along straight lines. Any two geodesics on a plane meet in a point, but if they are parallel they do not meet at all.

Geodesics on the surface of a sphere are along the greater circles. On a sphere two geodesics, even if they appear parallel, always intersect in two points.

In the case of our Earth all longitudes are geodesics. At the Equator all longitudes appear parallel to one another, but they intersect at the poles.

Parallel geodesics on the surface of a pseudo-sphere approach one another as close as possible, but never intersect.

37. What determines the nature of geodesics ?

The nature of the geodesics on a surface depends upon the curvature of the surface.

A plane surface has zero curvature.

A sphere has constant positive curvature everywhere on its surface.

The surface of an egg has positive curvature but it varies from place to place.

A pseudo-sphere has constant negative curvature.

A surface like a saddle has negative curvature.

38. Must a "straight line" extend to infinitly in both directions ?

Parallel lines in Euclidean geometry do not intersect and no matter how far extended on either side always remain a constant distance apart. A straight line is therefore supposed to extend to infinity in both directions.

Riemann suggested that there was no logical necessity for such a notion and that all straight lines if sufficiently extended might return to themselves and be of the same length like longitudes on the surface of the Earth.

In the case of a sphere like the Earth every longitude intersects every other longitude in two points, namely the North and the South pole so that every pair of "straight lines" always intersects and encloses an area, and no two "straight lines" can be parallel.

39. But how could a straight line obey both Euclid and Riemann ?

Euclid's tacit assumption implies that a straight line may be extended infinitely. According to Riemann a straight line, if sufficiently extended, might return to itself.

The apparent conflict is resolved by Riemann pointing out an important distinction between infinite and unbounded.

The straight line may be unbounded and not infinite just as the surface of a sphere is unbounded but not infinite. Such a straight line without sacrificing demands of consistency serves both Eucid and Riemann perfectly well.

40. Which is the geometry of the Earth ?

For all *ordinary* purposes the surface of the Earth behaves as if it were flat. Thus for constructing buildings, bridges, tunnels, roads, etc. and fields for games and sports, the shortest distance between two points is a straight line and the sum of the three angles of a triangle is two right angles, and the geometry applicable is Euclidean.

41. What about large distances on the Earth ?

But consider a large triangle on the surface of the Earth formed by an arc of the Equator and the parts of two longitudes i.e. great circles drawn from the North pole to the ends of this arc. See figure below :

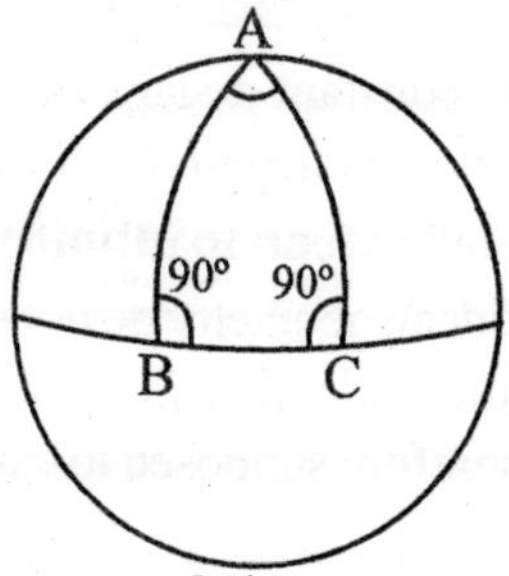

Sphere

∠A + ∠B + ∠C

Greater than 180°

The two base angles are each equal to 90° so that all the three angles of the triangle would together add up to more than 180°.

The shortest distance between any two points is no more a straight line but a portion of the arc of the longitude, so that Euclidean geometry ceases to hold good.

In fact even when two points on the surface of the Earth are more than a few hundred feet apart, allowance has to be made for the curvature of the Earth for determining exact distance between them.

42. How flat or how curved is the Earth ?

A straight line in a plane is straight and is said to have no curvature, whereas in the case of a circle, the smaller the circle is, the more curvature it has.

If we take a circle of one foot radius to have unit curvature, the curvature of a circle of one yard radius will be equal to one third unit; and on this scale the curvature of a great circle on the surface of the Earth will be about one part in 21 million. This curvature is so small that an arc of such a circle is practically indistinguishable from a straight line.

The geometry of the Earth is, therefore, Euclidean for small lengths or distances but not for large ones.

43. What is the geometry of the space we live in ?

Gauss, 'the prince of mathematicians', took three distant mountain peaks to form a triangle and found that the sum of the angles of the triangle thus formed was 180° within the limits of experimental error.

The experiment proved inconclusive because though the triangle he used was big enough compared with ordinary diagrams drawn on paper, yet it was very small compared with the dimensions of the universe.

If in stead of three distant mountain peaks, we took three distant stars, the experiment would still not be conclusive, though this time for entirely different reasons.

44. What are these reasons ?

Since in this case, the measurement of angles will be by means of rays of light, and these during their voyage through space would bend in accordance with the intensity of the gravitational field they pass through, the result of measurement will tell us more about the laws of propagation of light than about the nature of space, Euclidean or otherwise.

45. What precisely in meant by space ?

One view of space could be an absolutely empty space, void of all traces of matter, but in such a space nothing would distinguish a place or a direction, so that there are no places, no directions, and the absolutely empty space, therefore, reduces to nothing more than an abstraction.

46. What is the other view ?

The other view of space is that "space is the form of existence of matter," so that properties of real space are properties of certain relations of material bodies, for example, their dimensions, mutual positions etc.

According to this view real space is inseparably connected with matter. Matter determines the geometry and geometry accounts for the phenomenon formerly attributed to gravitation.

Not only this, but as Einstein has shown, space cannot be separated from time, and together they form a single form of existence of matter, namely space-time.

47. What if space and time are treated as distinct entities ?

The structure of space-time is complicated and space cannot be separated from time except under certain assumptions, in which case space turns out to be Euclidean in domains small compared with cosmic dimensions, but in large domains containing big masses of matter, the deviation from Euclidean geometry becomes apparent.

48. What approximately is the geometry of the universe ?

Many hypotheses have been put up on the structure of the universe as a whole assuming the distribution of mass to be uniform and the universe non-static.

These assumptions have been made to simplify matters and enable us to have an approximate idea of the actual scheme of things.

Under such assumptions a hypothetical theory proposed by the soviet physicist Friedmann shows the geometry of the space on the whole to be Lobachewskian.

49. Which geometry applies to elementary particles ?

Just as Euclidean Geometry does not hold good for large distances in space, it does not hold good for extremely small distances either.

Non-Euclidean geometry is applicable to distances within and between atoms, molecules, elementary particles, etc.

50. Are only three geometries possible ?

It would appear that an infinite number of geometries is possible, because starting with any postulates whatever a geometry can be built up, provided only that the postulates do not lead to any contradiction.

A surface can often be found to which a new geometry will be applicable.

Alternatively, however complex a surface may be, a geometry peculiarly suited to it could be constructed.

51. What is Projective Geometry ?

Consider an artist standing before a scene he wants to paint. His canvas can be imagined to be a glass screen intercepted between the seene and his eye. The painting on the canvas thus turns out to be the projection of the scene on the glass screen with the centre of projection at the eye of the artist.

Since the real canvas is not transparent and the scene the artist wants to paint might exist only in his imagination, the artist needs a mathematical scheme to enable him to depict the 3-dimensional real world on a 2-dimensional canvas.

Projective geometry provides such a scheme. It deals with geometrical properties of figures that remain unchanged under such projections.

52. Which are such Projections ?

The most familiar example of such a projection is the shadow due to a point source of light.

Shadows of a circle due to a point source for different positions of the circle are not all circles. They are more or less flattened ellipses.

Shadow of a square can be rhombus, even some other quadrilateral.

The shadow of a right angle is not always a right angle.

Square floor tiles are not drawn square in a painting. But the impression created on the eye is the same as that of the original floor.

53. In what ways is the projection different from the original ?

In the projection duc to a point source, the size of the angles, areas and lines gets distorted, but there exist some properties which are not destroyed so that the structure of the original can usually be recognised on the canvas.

54. What are such properties ?

The properties are rather simple :

Projection of a point is a point and that of a straight line is a straight line, i.e., a straight line will not become a curve. It follows that the projection of a triangle will always be a triangle and that of a quadrilateral will always remain a quadrilateral.

Three important properties follow from these simple ones :

(i) If a point lies on a straight line, then after projection the corresponding point will also lie on the corresponding line. This is called incidence.

(ii) If three or more points lie on a straight line, then their corresponding images will also lie on a straight line. This is called collinearity

(iii) If three or more straight lines pass through a point, their images will also pass through a point.

This is called concurrence.

55. Where is Projective Geometry applied ?

Projective geometry has applications in aerial photography, architechture and in the problems of perspective which are studied especially by the artists.

56. In what way is the Projective Geometry different from the Euclidean Geometry ?

Theorems of Euclidean Geometry deal with magnitudes of lengths, angles and areas with the related concepts of congruence and similarity.

These are called metric properties. They deal with magnitudes and remain unchanged under the class of rigid motions.

Projective Geometry deals with projective properties or properties which remain unchanged under projection, i.e., incidence, collinearity and concurrence.

57. Is the distiction between metric and projective properties of figures essential.

The distinction between metric and projective properties of figures was done away with by an English mathematician Cayley. He treated the whole matter algebraically and brought about a unification of the two.

58. What is coordinate Geometry ?

Coodinate gometry* is a study of geometry through algebraic methods.

It makes systematic use of the fact that there is a natural correspondence between the real numbers and the points in space.

Take any point O on a staight line. Let it be called the origin, i.e. the starting point for all measurements along the line. Then every real number corresponds to a point on the line and vice versa. The real number is called the coordinate of the point.

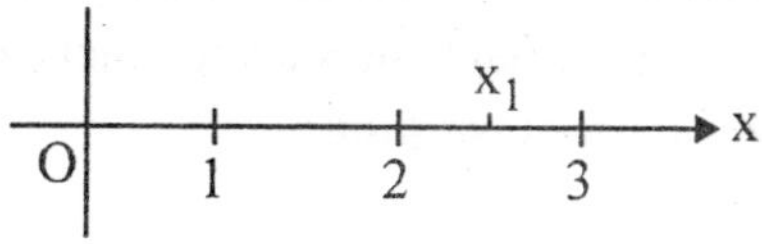

Again, take two perpendicular lines, ox and oy, called the axes, through the origin O. Then the position of any point P in the plane is determined by its distances x_1 and y_1 from the vertical line oy and the horizontal line ox respectively in that order. The ordered pair (x_1, y_1) of real numbers defines the point P in the plane, and is called its coordinates.

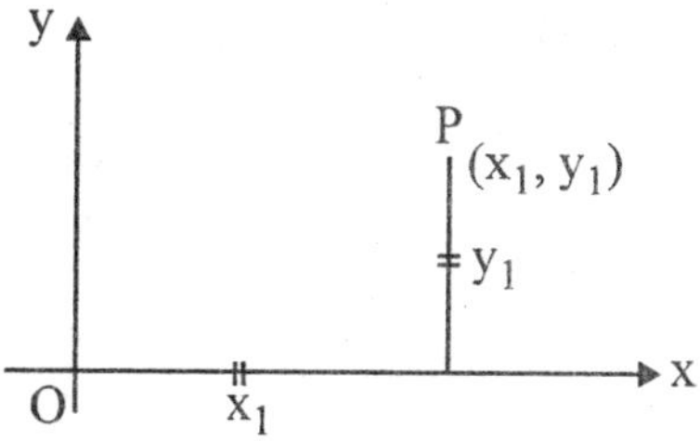

Coordinate geometry is also called Analytical Geometry or Cartesian Coordinate Geometry in honour of its inventor De cartes.

59. How is it a more powerful tool than ordinary geometry ?

The power of coordinate geometry lies in the fact that it investigates geometric objects by algebraic methods.

The concept of coordinates transforms problems of geometry into those of calculations with algebraical quantities.

And algebraical calculations are far easier to work out than manipulating geometrical proofs which involve lot of intuition and insight with figures and diagrams !

---

* It was developed chiefly by Rene' De Cartes, a French mathematician. He published his book in 1637.

Coordinate geometry has, therefore, been rightly credited with "having emancipated geometry from the slavery of diagrams".

60. How is it done ?

By the coordinate method simple algebraic equations of the first degree in two unknowns x and y find a pictorial meaning and represent straight lines so that the study of the geometrical object called straight line is done through the study of such equations. This method is easier and for more rewarding !

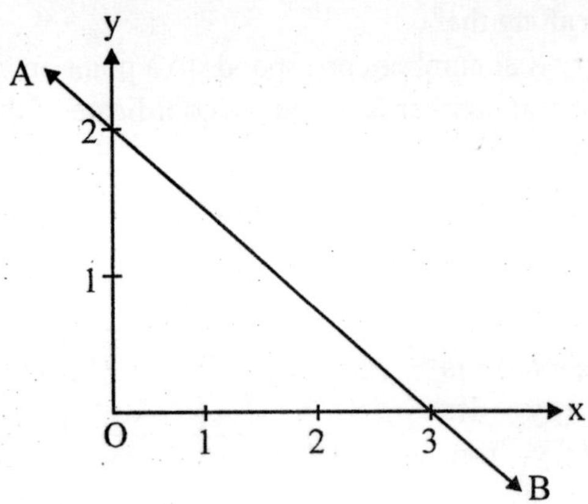

The equation $2x + 3y = 6$,

or which is the same as $\frac{x}{3} + \frac{y}{2} = 1$, obtained on dividing both sides by 6,

is pictorially represented by the straight line AB.

Likewise, the general equation of a straight line is taken as

$$ax + by + c = 0$$

Algebraic equations of the second degree in two unknowns x and y represent curves in a plane.

61. Which are these curves ?

The most familiar of such curves are the circle, the parabola, the ellipse, and the hyperbola. A pictorial representation of each with its equation is given below :

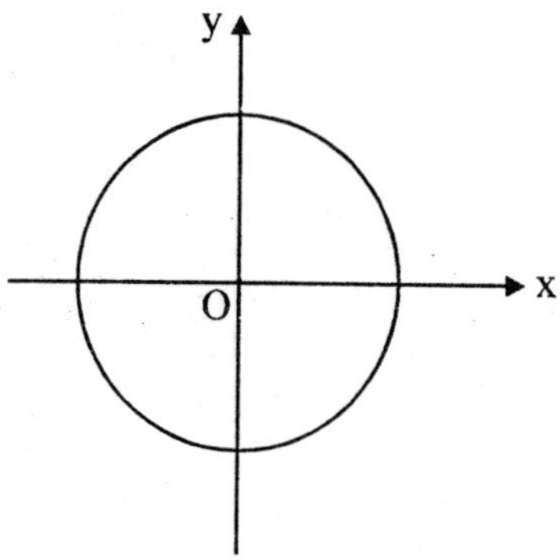

Circle : $x^2 + y^2 = a^2$

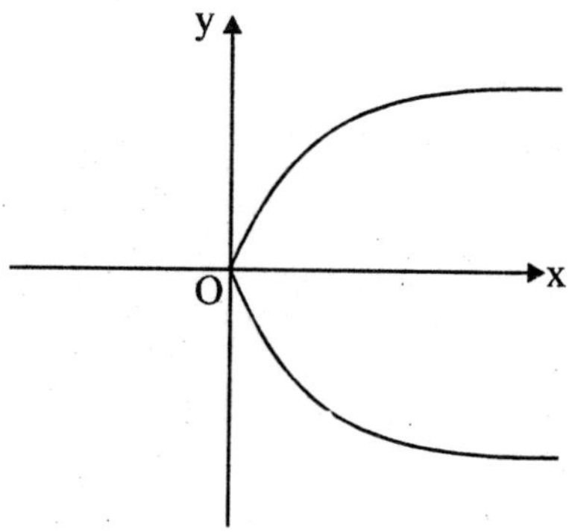

Parabola : $y^2 = 4ax$

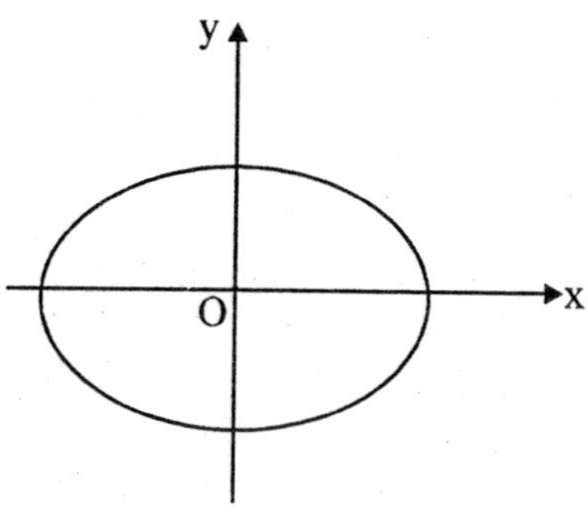

Ellipse : $\frac{x^2}{a^2} + \frac{y^2}{b^2} = 1$

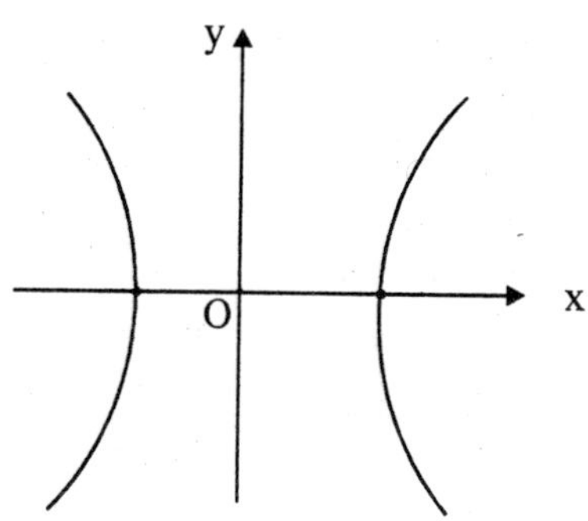

Hyperbola : $\frac{x^2}{a^2} - \frac{y^2}{b^2} = 1$

62. What are conic sections ?

Sections of a cone by different planes are called conic sections.

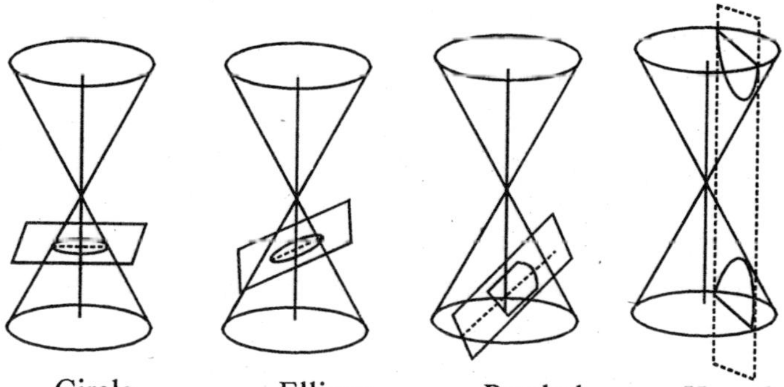

Circle Ellipse Parabola Hyperbola

If a plane cuts the cone perpendicular to its axis, the section is a circle.

If the intersecting plane is inclined to the axis of the cone, an ellipse results.

If the intersecting plane is parallel to a generator of the cone, a parabola is obtained.

If the intersecting plane cuts the double cone, two branches of a hyperbola are obtained.

If the intersecting plane cuts the double cone and also passess through the vertex, a pair of straight lines through the vertex is obtained.

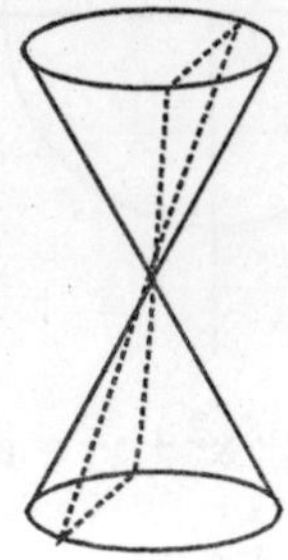

Pair of straight lines

63. What is meant by the reflection property of the parabola ?

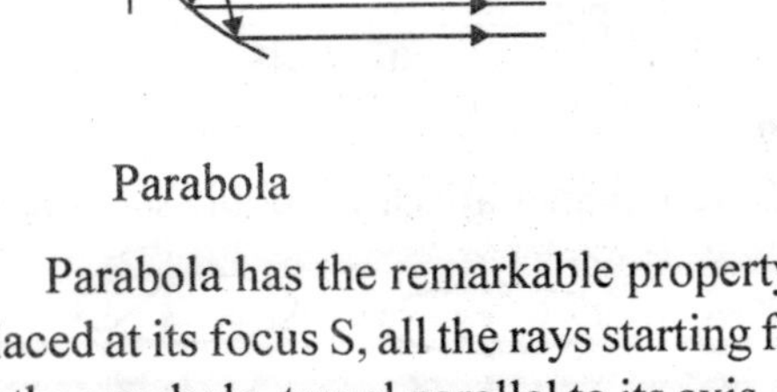

Parabola

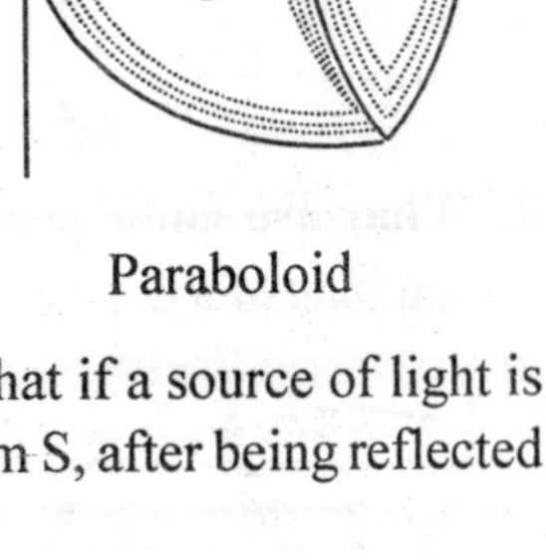

Paraboloid

Parabola has the remarkable property that if a source of light is placed at its focus S, all the rays starting from S, after being reflected at the parabola, travel parallel to its axis.

This is known as the reflection property of the parabola.

It is on account of this property that the reflectors placed behind the automobile head-lights are made of paraboloidal shape, i.e., of the shape generated by revolving the parabola about its axis.

This renders it possible to achieve visibility at a much longer distance along the road.

64. What is the acoustical property of the parabola ?

Rays starting from the focus are reflected parallel to the axis of the parabola.

Conversely rays coming parallel to the axis are reflected and brought together at the focus.

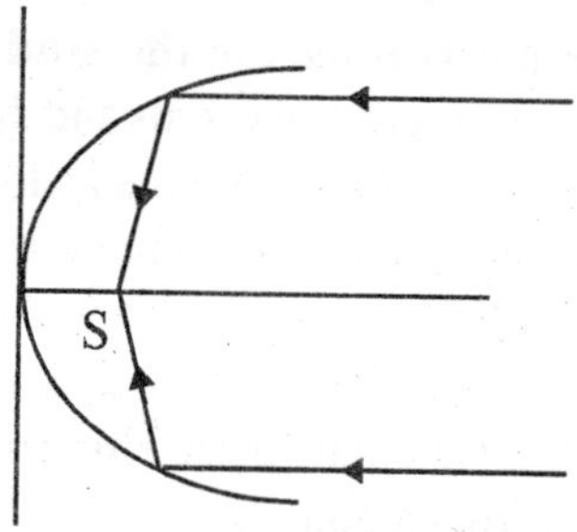

Since sound waves also behave in like manner, the property of sound converging at the focus is known as the acoustical property of the parabola.

This explains why in some art galleries words whispered at some distance are clearly audible to an ear held at a certain spot, but not elsewhere.

The spot, S, is actually the focus of the parabolic structure.

65. What is the reflection property of the ellipse ?

Ellipse has the property that rays starting from any one focus, say S_1, after reflection at the ellipse pass through the other focus S_2.

This is known as the reflection property of the ellipse.

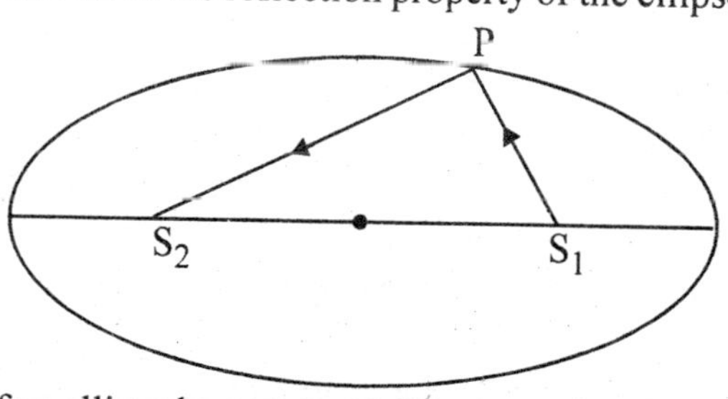

Thus, if an ellipse be constructed from a strip of polished metal, rays of light starting from one focus will come together at the other focus.

An object placed at S_2 will become illuminated due to a source of light at S_1, though S_1 and S_2 be quite far apart.

66. What is the acoustical property of the ellipse ?

Reflection of sound from one focus to another via ellipse is known as the acoustical property of the ellipse.

This explains why in some elliptical art galleries observers standing at two particular spots can hear each other's whisphers without being audible to the people in between.

67. What was the motivation for the study of these curves ?

It was motivated by a series of events and discoveries and pressing needs, quite important of which were the following :

Kepler had discovered that planets move around the Sun in ellipses and Galileo* discovered that a stone thrown into the air traces out a parabola. Likewise bullets ejected from guns also trace out parabolas.

It was necessary to calculate these ellipses and also the parabolas described by bullets from a gun.

68. What were other needs ?

Astronomy with stationary Earth had been invalidated, and so was the classical Greek mechanics. These needed re-view and re-examination.

The rapid development of long-range nagivation called for maps that would correlate paths on the globe with paths on flat maps.

Other regions of natural science had also like problems awaiting exact formulation and solution.

69. Why were the properties of conics not utilized, when they were well known even to the ancient Greeks ?

Properties of conics were well known to the ancient Greeks about 2000 years before Des Cartes, but they formed only a part of geometry. No method of using them in other areas was known. The device of coordinate system replaced curves by equations, which are relatively very easy to handle. And the technique opened the gateway to a vast treasure-house never dreamed of before !

70. Was the algebraic technique sufficient for working with curves ?

No, it was soon found that these techniques could not cope with slope and curvature, which are fundamental properties of curves.

---

* Kepler Published his findings in the year 1609. Galileo put up his ideas in a book, which came out in 1638.

71. How are slope and curvature defined ?

Slope is the rate at which a curve rises or falls per horizontal unit.

Curvature is the rate at which the direction of the curve changes per unit along the curve.

Slope of a straight line remains constant throughout its length, and its curvature is zero.

Curvature of a circle remains constant all along its length.

The slope and curvature both change from point to point for all other curves.

72. How are slope and curvature calculated ?

Differential calculus provides the methods for calculating these quantities for different curves.

73. What is Differential Geometry ?

The study of curves and surfaces with the aid of differential calculus is known as Differential Geometry.

It deals with a variety of problems beyond the calculation of slope and curvature.

It also deals with the very important problem of geodesics, i.e., the determination of the shortest distance between two points on a surface.

74. What is the coordinate geometry of three dimensions ?

If we add one more line oz perpedicular to both ox and oy, i.e., perpendicular to the plane of the paper and measure distances parallel to ox, oy and oz in that order, a point P in space can be defined by the ordered triplet (x_1, y_1, z_1) of real numbers.

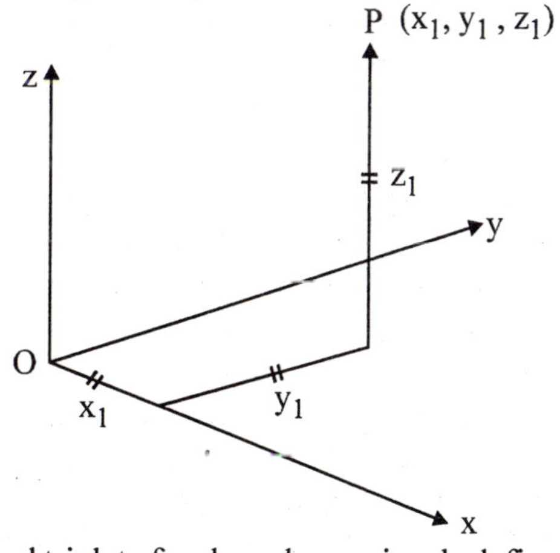

Conversely, any ordered triplet of real numbers uniquely defines a point in space. (x_1, y_1, z_1) are called the coordinates of the point P.

Coordinate geometry of three dimensions deals with points in space or which is the same thing, ordered triplets.

75. What is the geometry of n dimensions ?

Cayley and the German mathematician Grassmann independently generalized the coordinate geometry of two dimensions.

In the coordinate geometry of two dimensions, a point is determined by two coordinates and the distance between two points having coordinates (x_1, y_1) and (x_2, y_2) is given by

(x_1, y_1) and (x_2, y_2) is given by

$$\sqrt{\{(x_2 - x_1)^2 + (y_2 - y_1)^2\}}\ .$$

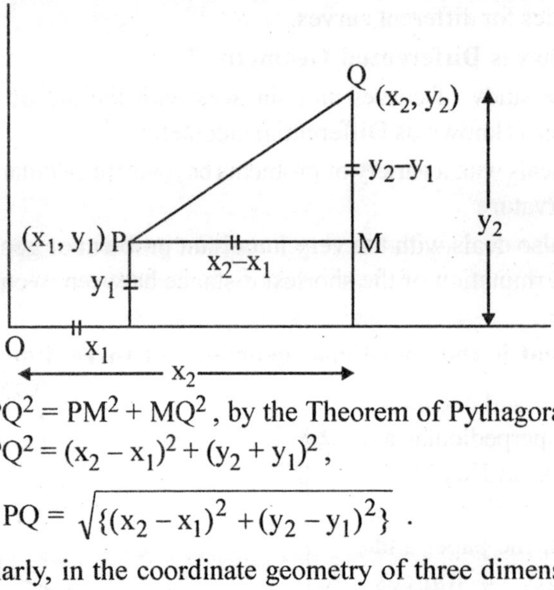

$PQ^2 = PM^2 + MQ^2$, by the Theorem of Pythagoras,

or $PQ^2 = (x_2 - x_1)^2 + (y_2 + y_1)^2$,

or $PQ = \sqrt{\{(x_2 - x_1)^2 + (y_2 - y_1)^2\}}$.

Similarly, in the coordinate geometry of three dimensions, the distance between two points having coordinates (x_1, y_1, z_1) and (x_2, y_2, z_2) is given by

$$\sqrt{\{(x_2 - x_1)^2 + (y_2 - y_1)^2 + (z_2 - z_1)^2\}}\ .$$

This can be generalized and in the coordinate geometry of four dimensions, the interval between two points having coordinates, $(x_1, y_1, z_1, t_1,)$ and (x_2, y_2, z_2, t_2) is given by

$$\sqrt{\{(x_2 - x_1)^2 + (y_2 - y_1)^2 + (z_2 - z_1)^2 + (t_2 - t_1)^2\}}\ .$$

It can be further generalized so that in the coordinate geometry of n dimensions, the interval between two points having coordinates $(x_1, x_2, x_3, x_4........., x_n)$ and $(y_1, y_2, y_3, y_4 ,....... y_n)$ is given by

$$\sqrt{\{(y_1 - x_1)^2 + (y_2 - x_2)^2 + (y_3 - x_3)^2 + (y_4 - x_4)^2 + ...(y_n - x_n)^2\}}\,.$$

Every concept in the two dimensional geometry can thus be generalized into the n-dimensional analogue. Since the space we live in is three dimensional, geometrical interpretation is not available beyond three dimensions but the analogy is extremely useful.

76. Of what use is the coordinate geometry of four dimensions ?

The coordinate geometry of four dimensions has been of immense use to the physicists.

Just as a point in a plane is completely determined by two numbers called coodinates and a point in space by three coordinates, an event is determined by three coordinates giving position in space and the fourth coordinate giving time of occurrence.

The distance between two events, i.e., the space -time interval, as it is usually called, is given by :

$$\sqrt{\{(x_2 - x_1)^2 + (y_2 - y_1)^2 + (z_2 - z_1)^2 + (t_2 - t_1)^2\}}.$$

This geometry has been used as an essential tool in the development of the Theory of Relativity and in the study of space, time and gravitation.

77. What is the concept of space in Mathematics ?

The term space assumes two meanings.

In one sense it is the ordinary real space, i.e. the space of our experience.

In the other sense it is the "abstract space", which is taken to be a collection of homogeneous objects in which spacelike relationships hold good. For example, "distance" between two objects can be defined in this space.

In mathematics it is always the second meaning that is of relevance.

78. What is a point ?

The concept of a point in two-dimensional coordinate geometry is that of an element of space whose position can be fixed up by two distances. Two dimensional space can, therefore, be looked upon as a collection of all such elements whose position can be fixed up by two lengths.

Similarly, three-dimensional space can be looked upon as a collection of all such elements whose position can be fixed up by three lengths.

With three coordinates the limit of visual interpretation is reached because it is not possible to comprehend in real space the position of a point with four or more coordinates.

79. What is the way out ?

In stead of assigning three lengths for fixing the position of a point in three dimensional space, let us say we assign three numbers to fix the point. The point, then, turns out to be a mere ordered triplet rendering it unnecessary to enquire as to where actually in space the eye will locate it ?

Once the compelling instinct to visualize the point has been overcome and the point identified as an aggregate of 3 numbers, there could no longer be any hesitation in replacing the number 3 by the general number n. And we have a "space" of n dimensions, where n can take up values greater than 3.

A "point" is, then, better termed an "element" and "space" a "manifold".

80. Is manifold a more general concept ?

The term "manifold" is more general and more accurate than the physical term "space".

A manifold roughly resembles a class.

A plane is thus a class composed of all those points that are uniquely determined by two coordinates, and is therefore a two-dimensional manifold.

Similarly the space of the three-dimensional coordinate geometry may be regarded as three dimensional manifold since three coordinates are required to fix points therein.

If n numbers or coordinates are required to fix up each member of a manifold, whether it be a space or any other class, it is called an n-dimensional manifold.

The manifold is supposed to have no attribute at all except that it is a class.

81. Do we have other manifolds ?

We have many kinds of manifolds that have nothing to do with either space or geometry. A three-dimensional manifold would be a class of elements, each of which would require exactly three numbers to specify it.

A group of people can be thought of as a manifold,– and a three-dimensional mainfold, it three numbers x_1, x_2, x_3 representing age, height and weight are necessary and sufficient to distinguish them.

The same group can be thought of as a four-dimensional manifold, if four numbers x_1, x_2, x_3, x_4 representing age, height, weight and house number are used. The group becomes a five-dimensional manifold if one more number x_5 representing income is taken in.

We can also think of a four-dimensional manifold of particles of a gas, using three dimensions to fix up their position and one dimension to fix their density.

82. What is the advantage of such representation ?

Suppose that we want to illustrate the dependence of the pressure of a gas on its volume.

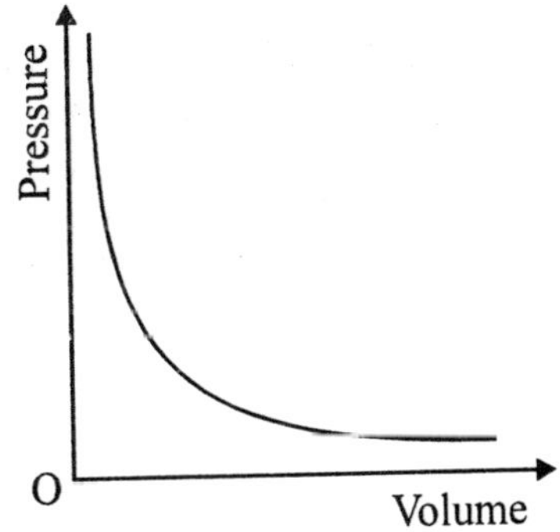

This can be done by taking two axes in a plane, representing volume along one axis and pressure along the other. The resulting curve would be a hyperbola for an ideal gas at constant temperature.

But if we have a more complicated system whose state is given not by two attributes but say five, then the graphical representation of its behaviour involves a five-dimensional space, i.e., the state of this system can be thought of as a point in some five-dimensional space.

In the same manner, if the state of a system is given by n attributes or variables, its state can be thought of as a point in some n-dimensional space.

The advantage of such representation is that a study of the system is rendered possible by the application and extension of the usual geometrical analogies and concepts.

83. Is our real space embedded in a four-dimensional space ?

The concept of the fourth dimension is only an abstract concept created to describe in geometric language those ideas that cannot be described by usual geometric representations.

It was developed to meet the demands of systems that depend upon several variables. But it was intended only as a mathematical method of modelling physical phenomenon and has nothing to do with the nature of actual space, which in scientific fiction is often described as four dimensional.

The idea that our three-dimensional space is embedded in an actual four dimensional space is a piece of mystic thinking and only a distortion of scientific concepts.

84. Can geometric concepts be also applied to algebra ?

Problems of algebra which involve only two or three unknowns often have geometrical interpretation. This means that if a problem has an easy or obvious solution from the geometrical point of view, that solution also serves for the problem taken algebraically.

An example will illustrate the point.

Suppose we want to know the integral solutions of the inequality $x^2 + y^2 < N$.

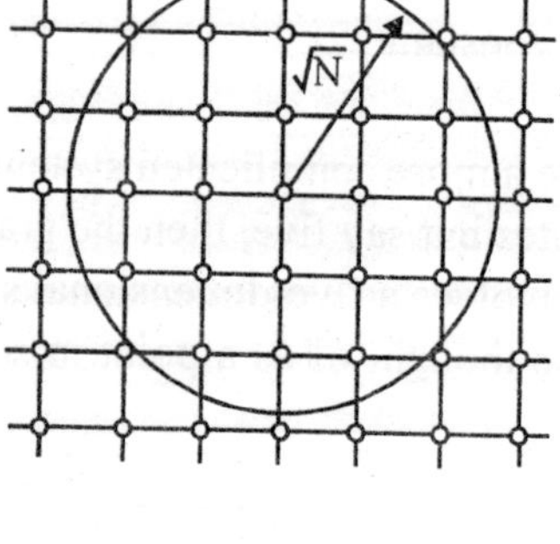

Geometrically the inequality $x^2 + y^2 \angle N$ represents the interior of the circle whose centre is at the origin and radius equal to $\sqrt{N}$, and the problem reduces to this :

How many points with integer coordinates lie inside the circle of radius $\sqrt{N}$?

Such points are the vertices of the squares with sides of unit length inside the circle. The number of such points inside the circle is approximately equal to the number of squares within the circle, which is equal to the area of the circleof radius $\sqrt{N}$.

Therefore, the number of integer solutions of the inequality is about πN.

The error in the result tends to zero for large valus of N.

It is evident that the solution though obvious geometrically is not so from the point of view of algebra.

85. Does the result have an analogue in higher dimensions ?

The corresponding problem in three variables can be similarly solved, but if the number of variables is increased beyond three, the method ceases to be applicable.

The result can however be generalized to any number of variable so that the corresponding problem in n variables has a solution in algebra though geometrical interpretation is no longer available since our real space is only three dimensional.

86. What is the geometry of colour space ?

Space is looked at as a collection of points. But if the "points" be objects, events or states themselves, then this collection can be regarded as a "space" of its own kind.

The terms point, straight line, distance etc. are then, used with a highly modified significance.

An example of such a space is the space of colours.

87. How does this space respond to geometry ?

Normal human vision is supposed to be three-coloured. The perception of a colour C is a combination of three basic perceptions : red R, green G and blue B with varying intensities, so that we have

$$C = xR + yG + zB,$$

where x, y, z denote intensities in certain units.

A point can be moved in space left and right, back and forth, up and down, so a perception of colour can be changed continuously in the three directions by changing its constituents R, G and B.

The set of all possible colours can thus be looked upon as the 3-dimensional colour space.

Since the intensities cannot be negative, x, y, z are always positive. When $x = 0$, $y = 0$, $z = 0$ we have a complete absence of colour.

88. How are point, segment and distance defined in this space ?

Here, a "point" is a colour, the "segment" AB is the set obtained by mixing the colours A and B. The "distance" between two colours is defined as the length of the shortest line joining them. The measurement of length and distance in the colour space is, therefore, defined by a certain non-Euclidean geometry.

89. Does the geometry of colour space have any applications ?

The geometry of colour space provides an accurate mathematical basis for solving problems on dyes in the textile industry, on the difference of colour signals, and related fields.

90. What is finite geometry ?

Our concept of space is that of an aggregate of points or elements, which are infinite in number. But we can also have a geometry of only a finite number of points, say twenty five or so.

The terms point, straight line, distance, parallel etc. are used with meanings suitable to the system under study.

Such a finite geometry applies to certain problems, and algebra and number theory; and has been found useful in coding theory and in the construction of experimental designs.

91. What is Topology ?

It is a new development in geometry almost exclusively of the twentieth century and is one of the most sophisticated and vigorous branches of modern mathematics.

It is a type of geometry which studies properties of figures and surfaces that remain unchanged during stretching, bending, contracting, and twisting operations.

92. In what way is it different from other geometries ?

Unlike other geometries, it does not deal with magnitudes of lengths and angles, and is a non-quantitative geometry.

It deals with relations dependent upon position only. In other words, it deals only with the topological properties of figures and surfaces.

93. What are topological properties of figures ?

These are those properties of figures that persist even when the figures are subjected to deformations so drastic that all their metric and projective properties are lost.

Consider a circle (i.e., the curve only and not the enclosed area) drawn on a sheet of rubber. By stretching, contracting, bending, twisting, but not tearing, fusing or overlapping it can be deformed into an ellipse, a triangle, a square, or any other regular or irregular shape.

Every such transformation is called a topological transformation. Its distinguishing property is that the parts of the figure that are in contact remain in contact, and the parts that are not in contact cannot come in contact. In brief, in a topological transformation neither breaks nor fusions can arise.

Under such operations, properties like distances, angles and areas undergo change and are not topological properties.

94. Inside and Outside ! Are these topological properties ?

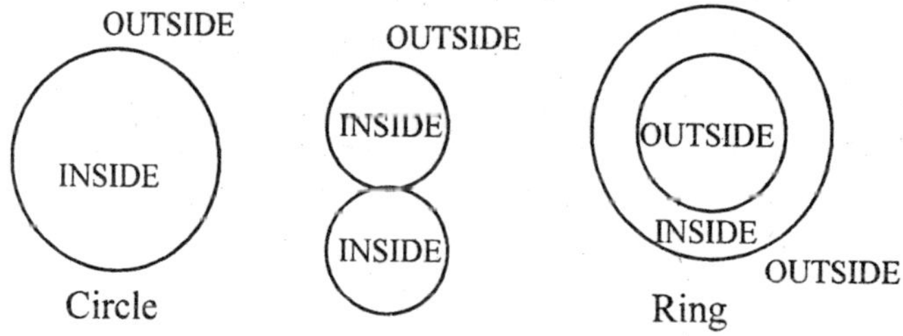

Figure like 8

The fact that the circle has one "inside" and one "outside" is a topological property.

The figure like 8 has two loops and therefore has two "inside" and is topologically not equivalent to a circle or a triangle, each of which has only one "inside".

A ring bounded by two concentric circles has two "outside" and one "inside."

95. What are topological properties of surfaces ?

Take the surface of a sphere. It has two properties which are both preserved under an arbitrary topological transformation.

First, the surface of a sphere is closed. By this is meant that unlike cylinder it has no edges—a cylinder is bound by two edges.

Second, every closed curve on the sphere splits the surface into two dis-connected parts.

An inflated inner tube or a ring, called a torus, does not have this property. If a torus is cut perpendicular to its length, it is not split into two parts but is turned into a bent tube, which can be straightened into a cylinder by topological transformtion. Thus every closed curve on the torus does not split it into two parts.

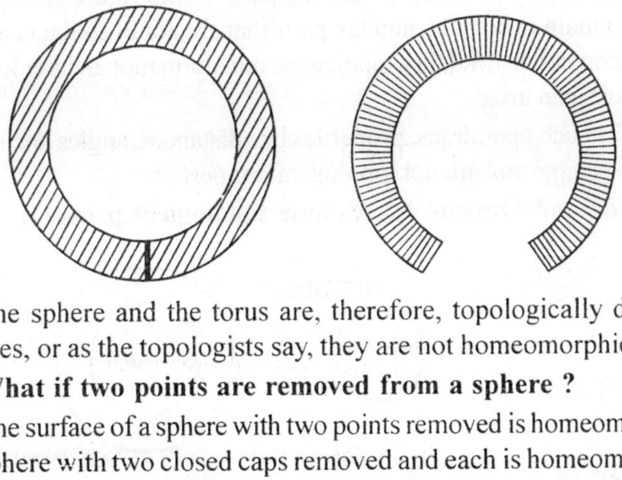

The sphere and the torus are, therefore, topologically distinct surfaces, or as the topologists say, they are not homeomorphic.

96. What if two points are removed from a sphere ?

The surface of a sphere with two points removed is homeomorphic to a sphere with two closed caps removed and each is homeomorphic to a cylinder. Spheres and cubes belong to the same topological type, i.e. they are homeomorphic.

97. What about a pair of gloves ?

Take a pair of gloves. One is a right-handed glove and the other a left-handed one. If the right-handed glove is turned inside out, it becomes left-handed. The left handed one becomes right-handed if turned inside out. Topologcial reasoning enables us to predict this change of form.

98. What are the fundamental concepts of topology ?

The concept of adjacentness, neighbourhood, infinite nearness and the concept of dissection of a body (division into parts) are fundamental concepts of topology.

Some allied concepts are insideness and outsideness, right-handednness are left handedness, connectedness and unconectedness, and continuity and discontinuity.

99. Does topology deal with surfaces only ?

No, study of surfaces is just one domain. Topology has many aspects, but it can be broadly divided into three branches :

Combinatorial topology,

Algebraic topology,

Point-set topology.

The division is largely a matter of convenience rather than of logic, because there is considerable overlapping among the branches.

100. What is Combinatorial topology ?

It is the study of those intrinsic aspects of geometric forms that remain unchanged under topological transformations.

It treats figures as combinations of simple figures joined together in a regular manner in constrast to point-set topology which considers figures as sets of points.

101. What is Algebraic topology ?

Initially, topology developed as a study of surfaces. But it was soon found that its concepts were very initimately related to several problems of fundamental importance in various domains of mathematics. Algebraic methods, especially group theory, proved to be of immerse use in such investigations.

This algebraic machinary is called algebraic topology and is a powerful tool for proving topological results.

In also yields lot of results in higher dimensions, where we cannot see, but can only reason.

102. What is Point-set topology ?

While topology was being developed as a study of surfaces, it was also realised that topology of elementary surfaces alone was insufficient and that the solution of problems in one, two, three and n-dimensional topology was necessary. These studies were found to involve set-theoretical considerations and developed into what is called Point-set topology or General topology.

The class of geometric figures under consideration in this topology is extremely wide. A point in this topology can represent a point of an ordinary geometric figure, a complete figure itself, or a whole system of geometry.

103. Why is topology called rubber-sheet geometry ?

One aspect of topology deals with deformation of figures without tearing or fusing their points. Since such deformations can be carried out over figures drawn on a rubber sheet, topology is sometimes referred to as rubber-sheet geometry.

But modern topology extends far beyond this introductory aspect.

104. Is contemporary topology a study in geometry ?

In its early stage topology was regarded as the "science of position" as it literally means, but it has since far outgrown its initial scope.

Regarding its change of character it has been well observed that "topology began as much geometry and little algebra, but now it is much algebra and little geometry."

Historically speaking, topology has developed along two distinct directions. In one the inspiration seems to have come from geometry, while in the other analysis has exercised the main influence.

105. Is it true to say that Topology is a study of continuity ?

It is now generally accepted that topology is the study of continuity.

But most important of all, it has come up as a subject that attempts to unite almost the whole of mathematics somewhat similar to philosophy seeking to coordinate all knowledge.

Today, topology has so deeply pervaded mathematics that it is now an indispensable tool of the modern mathematician, pure or applied.

106. What is meant by saying that Topology is the mathematics of the possible ?

This has reference to many questions which remain unanswered in other areas of mathematics but are decided by applying topological concepts.

For example, topology decides whether for certain problems solutions do exist or do not exist, though it does not usually tell how to find the solutions.

Similarly, it can tell whether certain conditions are possible or impossible.

107. Any specific example ?

Take an instance from Algebra. What is known as "The fundamental Theorem of Algebra" states that

Every algebraic equation of any degree n with real or complex coefficients,

$$x^n + a_1 x^{n-1} + a_2 x^{n-2} + \dots\dots\dots + a_n = 0,$$

has solutions in the field of complex numbers.

Here is a purely algebraic situation, i.e., an equation has roots or not, but there exists no purely algebraic proof of this important result. All proofs require a knowledge of the calculus of functions of several real variables, or of complex analysis.

But since the ideas and methods of topology have transformed large parts of these branches almost beyond recognition, it is generally believed that the theorem depends essentially on topological considerations.

108. Any other example ?

Again, take an instance from Differential Equations. Most physical phenomenon and problems of modern technology can be described mathematically by differential equations, i.e., equations involving rates of change. In these studies non-linear differential equations are of frequent occurrence but they are extremely difficult to solve. Topology can show what types of solutions of certain non-linear differential equations are possible, though here again the answers are qualitative and not quantitative.

It is in such context that topology is described as the mathematics of the possible.

109. Do topological ideas have any practical applications ?

Topological concepts are made use of in the design of networks meant for the distribution of electricity, gas and water, and for designing industrial automation.

They are made use of in the control of automobile traffic and guided missiles.

They are also applied to the designing of geographical maps.

The theory of dynamical systems has flourished due to topological ideas and concepts.

Modern function theory and symbolic logic are intimately related to topology.

110. What is a one-sided surface ?

Take a strip of paper and paste the ends together, A coinciding with C, and B with D. This makes a cylinder.

Strip of paper

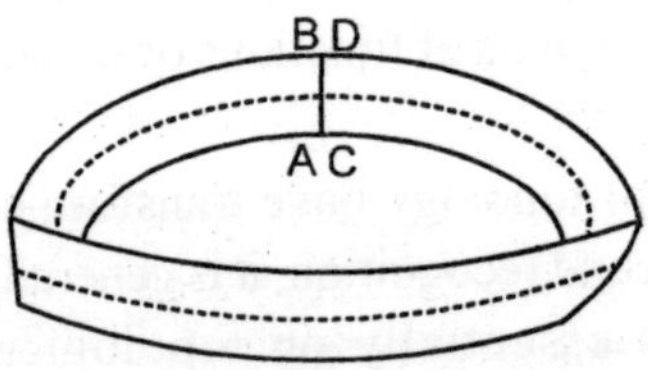

Cylinder

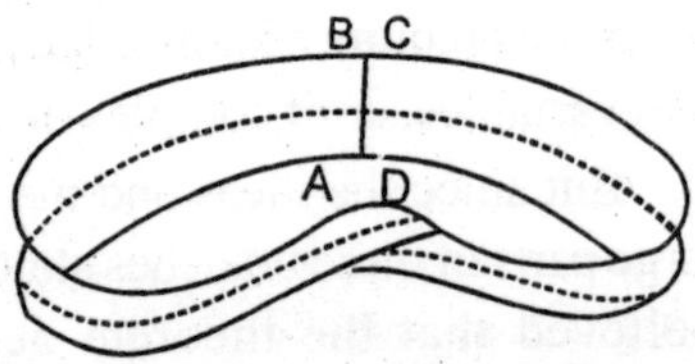

Mobius strip

The cylinder has two surfaces, the inner and the outer—one of them can be painted say blue and the other red.

Also, it has two edges, the upper and the lower. Now take another strip, give it half a twist and paste the ends together so that this time A coincides with D, and B with C. This is the famous Mobius strip, discovered by the German mathematician A. F. Mobius in 1858.

If we try to paint the sides of this object in two colours, we will find that it is imposible to do so, because it has only one surface !

Seems incredible but is worth verifying with a strip of paper or a piece of ribbon.

111. What is the moral of this experiment ?

It shows that even the apparently simple and sincere assertion that every surface has two sides can be wrong ! It is, therefore, that in mathematics rigorous logical proofs are demanded, howsoever obvious the assertions might seem.

112. Any other property of the Mobius strip ?

Another striking property of this surface is that it has only one edge, a single closed line ! If a driving belt is given half a twist before its ends are sewn together, it becomes a model of a Mobius strip. Such a belt may last longer because it wears equally on both sides due to friction over the wheels. In fact, it has only one side and one edge.

Again, if we cut a cylinder perpendicular to its axis along the middle line, we get two cylinders. But it we cut Mobius strip along the middle line, it still remains one piece ! It is interesting to verify it with a strip of paper.

113. What is Euler's formula for solids ?

The formula gives a relation between the vertices, edges and the faces of a simple solid.

It states that for any simple polyhedron,

$$V - E + F = 2,$$

where V is the number of vertices, E the number of edges, and F the number of faces.

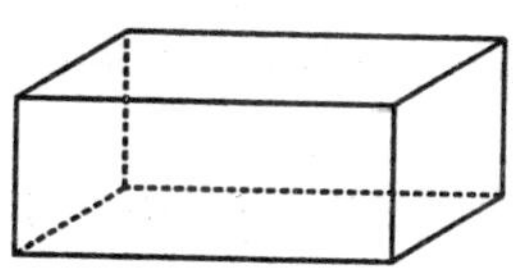

Parallelopiped

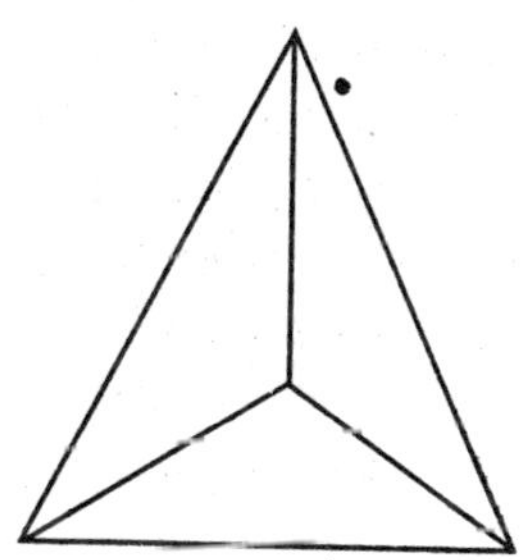

Tetrahedron

To illustrate, a parallclopiped or a cube has 8 vertices, 12 edges, and 6 faces, so that $8 - 12 + 6 = 2$ makes a valid statement.

Similarly, a tetrahedron has 4 vertices, 6 edges, and 4 faces, so that $4 - 6 + 4 = 2$ again makes a valid statement.

114. But how is this formula a result in topology ?

The formula remains valid even when the solid is subjected to all types of topological deformations, wherein, in general, the edges cease to be rectilinear, and the faces cease to be plane and change into curved surfaces.

This formula is therefore, supposed to be historically the first theorem in topology.

It was known to Des Cartes at about 1640, but was re-discovered by Euler in 1752.

115. What is the problem of the Seven Bridges of Koenigsberg ?

The German* city of Koenigsberg has its centre on an island in the river Pregel. In the seventeenth century this island was linked to the banks of the river by two bridges to each bank. The island was also linked by a bridge to another nearby island, which was also linked to each bank of the river by a bridge.

A schematic diagram is given below :

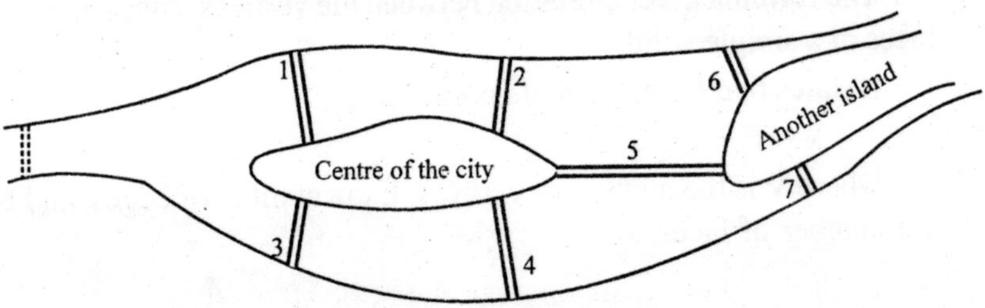

What puzzled the citizens was this :

How to plan a journey across all the seven bridges without crossing any one of them twice ?

This is known as the problem of the Seven Bridges of Koenigsberg.

116. How can such a journey be planned ?

Such a journey cannot be planned. Repeated trials show that the feat is impossible, but Euler worked out the general principle which underlies such problems called network problems in topology.

117. What is that principle ?

Before the principle can be explained, a few ideas and terms are to be borne in mind.

For simplicity, land areas are replaced by points and the bridges by lines joining the points.

The points are called vertices. A vertex is odd or even, according as the number of paths leading from the vertex is odd or even.

---

* The city of Koenigsberg lately became part of U.S.S.R.

Euler discovered that :

(i) If *all* the vertices in a connected network are even, the network can be traversed in a single journey beginning and ending at the same vertex as in the following cases :

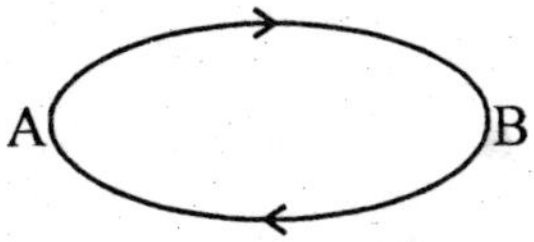

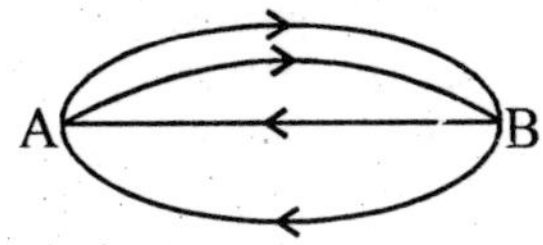

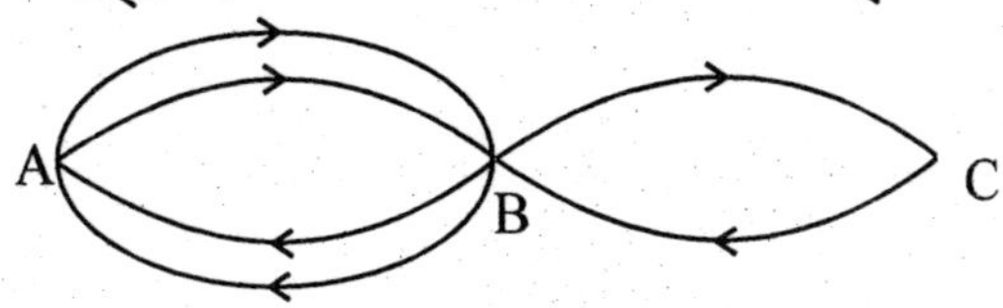

(ii) If the network contains two odd vertices, it can also be traversed in a single journey, but it is not possible to end at the starting point.

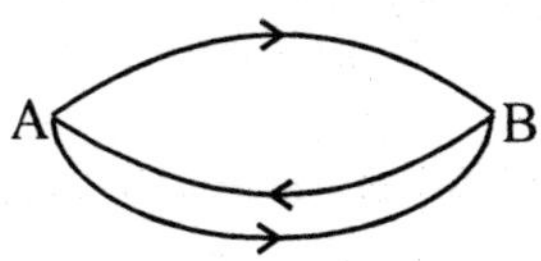

(iii) If the network contains 4, 6, or 8 odd vertices ; 2, 3 or 4 distinct journeys respectively will be required to traverse it. The number of journeys necessary to traverse a connected network is equal to half the number of odd vertices.

(iv) A network with an odd number of odd vertices is impossible to construct because each line is required to begin at a vertex and end at a vertex.

118. How is the principle applied to The Koenigsberg Bridges problem ?

In the present problem, when the two islands are replaced by the points A and B, the river-separated two land masses by the points C and D, and the seven bridges by seven arcs, the network assumes the following form :

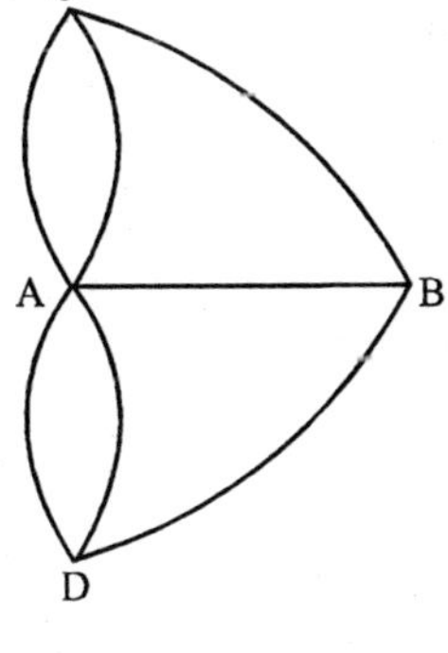

Since in this case all the four vertices are odd, two journeys are necessary to traverse the network. A single journey does not suffice.

119. What if one more bridge across the river is built ?

If an additional bridge be built, as has since been done across the river, shown in the schematic diagram by dotted lines at the extreme left, the network takes the following form :

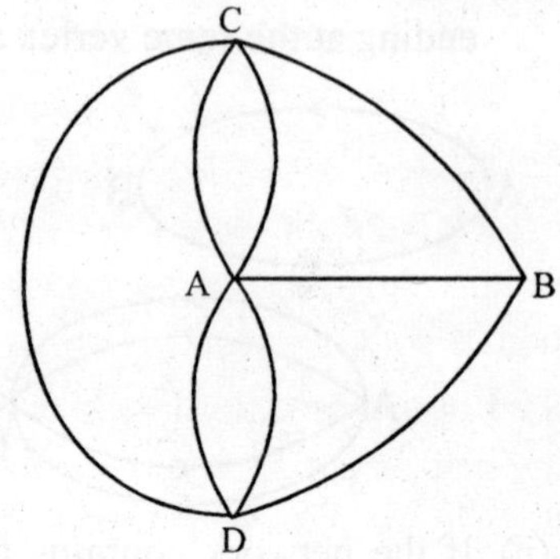

In this case, two vertices C and D are both even, and the other two, A and B, are both odd. A single journey will now suffice, but the journey must begin at one of the odd vertices A or B, and end at the other.

120. Why is the problem regarded so important ?

Because with the solution of the problem was born an entirely new branch of Mathematics.

When Euler presented the solution before the Russian Academy at St. Petersburg in 1736, not only was a long-standing problem settled but also with it was Topology founded.

121. What about taversing a star-like figure and a rectangle with diagonals ?

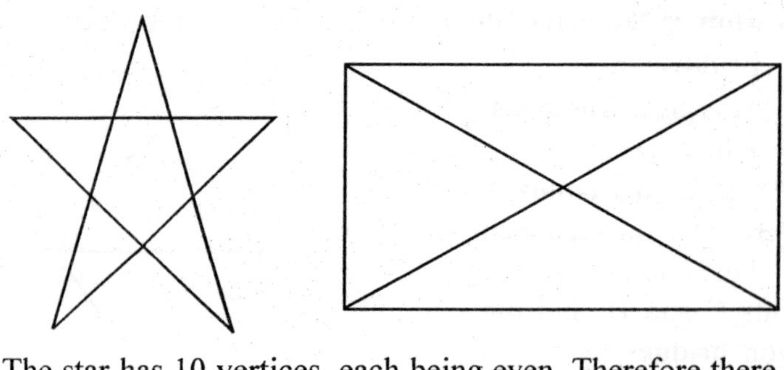

The star has 10 vertices, each being even. Therefore there can be no difficulty in traversing it in one stroke.

The rectangle with diagonals has 5 vertices, four of them odd, and one even. Two journeys are, therefore, required. It cannot be described in one stroke.

122. What is the four colour problem ?

While colouring a map, countries having common boundaries are to be coloured differently to distinguish them from one another.

It is common experience that four colours are sufficient to colour a map, no matter how many countries it contains or how complicated their boundaries are.

But to prove the fact that four colours are sufficient to colour any map in a plane or on a sphere has been found to be extermely elusive, and is known as the four colour problem.

123. Are three colours not enough ?

That less than four colours are not enough for all cases will be clear from the following map of four countries, where each touches the other three.

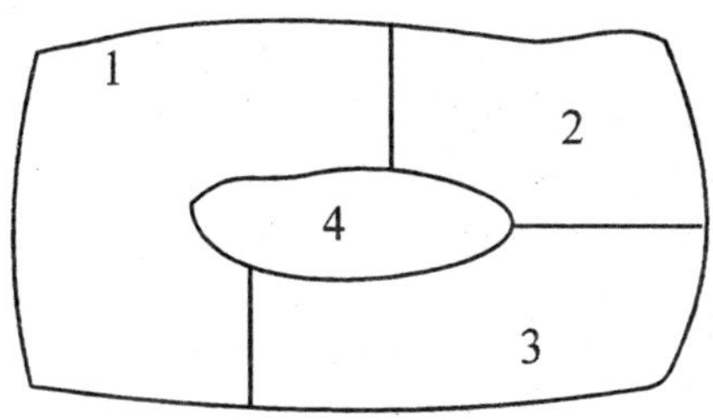

It is also true that no one has been able to produce a map whose colouring requires more than four colours.

124. How did it come up as a problem ?

It was first proposed as a problem by Mobius in 1840. Several gifted mathematicians tried their hand, but the solution eluded them for more than a century !

125. Who finally proved it ?

Only in 1976 Wolf Gang Haken and Kenneth Appel were able to prove the assertion, but the computer was also an almost equal partner in the development of the proof.

The proof runs into several pages and is extremely difficult.

126. What about maps drawn on the surface of a torus, i.e., an inflated inner tube ?

It had been proved that seven colours are necessary to colour any map on such a surface.

This implies that on such a surface maps can be constructed containing seven regions where each touches the other six !

127. How is it that geometric concepts are applicable to a variety of situations ?

Mathematics derives its creative power from intutions, of which geometry is a rich source—that explains why geometric concepts are applicable to a variety of situations.

Moreover, geometric methods and concepts preserve their advantages even in the abstract form.

Geometry supplies models not only of physical space but also of any structure whose concepts and properties fit the geometric framework.

128. Euclid again ! Why Euclid axiomatized geometry ?

Before Euclid, Geometry was just a collection of discrete unrelated results.

Euclid's aim was to so select a small number of initial assumptions or axioms that all what was known of geometry up of his time and also the geometrical truths yet to be discovered could all be deduced from them.

He axiomatized geometry to accomplish this remarkable task.

129. Is Euclid's work logically perfect ?

For more than two thousand years Euclid's "Elements" stood as a mathematical achievement of the highest order, but during the nineteenth century the standard of rigour in mathematical thinking grew higher, and logical gaps in Euclid's work began to be perceived.

There are many spots where the conclusions Euclid draws from his assumptions do not follow by logic alone.

130. Why were these logical gaps not noticed earlier ?

The reason why these gaps were not noticed by mathematicians for a very long time was that the figures which always accompanied Euclid's theorems made the assertions seem so obvious it never occurred to anyone that logically some of the propositions did not necessarily follow as they appeared to do. The figures unsuspectingly served to fill up the logical gaps.

It was, therefore, being increasingly felt that a stricter mode of presentation for geometry where proofs would be valid solely on account of their logical form, i.e. without any reference to the usual interpretation of geometrical concepts, should be developed.

131. What was done to achieve this end ?

The great German mathematician Hilbert worked out one such modern axiomatic treatment of Euclidean geometry.

He employed only three undefined terms—point, line and plane, and six undefined relations—on, in, between, congruent, parallel, and continuous, and twenty-one axioms.

He defined all other concepts of geometry, such as those of angle, triangle, circle, etc. in terms of these primitive terms or basic concepts.

132. Is Hilbert's the only possible axiomatic treatment of Euclidean geometry ?

No, there are many and can be many more. For example, a few years after Hilbert, Oswald Veblen worked out a different axiomatization using only the terms 'point' 'between', and 'congruent' with a set of axioms quite different from Hilbert's.

A still different axiomatization is due to E. V. Huntington, who used only two terms 'sphere' and 'includes' with obviously a still different set of axioms.

133. Is axiomatic method suitable for studies other than geometry ?

The impact of Euclid's axiomatic method was so great on the work of subsequent generations that it became a model for all rigorous demonstration in mathematics.

Consequently, during the nineteenth and early twentieth centuries many fields of enquiry which were so far developed in a more of less intuitive manner were put on axiomatic basis.

134. Does axiomatic method promote mathematical thinking ?

No, Axiomatic method can be looked upon as a mathematical activity based on preconceived ideas, whereas mathematics as a creative activity is developed independently of such ideas, therefore axiomatic method is unable to reveal the essence of mathematical thinking.

135. What, then, is the motive for axiomatizing other branches ?

The most powerful motive for axiomatizing various branches of mathematics has been the desire to set up a small but sufficient number of initial assumptions from which all the true statements in those fields can be deduced.

This method of axioms is now so completely accepted that one of the most striking features of the twentieth–century mathematics is its immensely increased use of the axiomatic approach in mathematical studies.

136. What was the result of this increased axiomatization of Mathematics ?

This increased axiomatization of abstactness of mathematics brought a serious difficulty over to the surface, that of consistency !

Since an axiomatic treatment must be consistent there should be a way of ascertaining that a given set of postulates underlying a new system is internally consistent so that no mutually contradictory theorems could be decuded from the set.

If the postulates are about a familiar domain of objects, it is always possible to test whether they are true of them, but in case of postulates about a new and an unfamiliar domain of objects, there appears no way of testing their consistency.

To illustrate, non-Euclidean geometries at a time when they were being developed and thought of as not representing any truth is the case in point.

There appeared no way of aswering the question : Is the Riemannian set of postulates consistent or that will it not lead to contradictory theorems ?

137. Where else does the problem of consistency arise ?

The problem of consistency also arises whenever a non-finite model is taken up for purposes of interpretation.

In case of finite models the consistency of the set can be ascrtained by inspection or enumeration but in case of non-finite models this is not possible.

And most of the postulate systems constituting the foundations of important branches of Mathematics can be satisfied only by non-finite models.

138. Could Hilbert succeed in establishing consistency of the Euclidean postulates ?

Hilbert undertook to interpret Euclidean postulates in the manner adopted in cartesian coordinate geometry so that they are transformed into algebraic truths. The consistency of the Euclidean postulates is thus established by showing that they are satisfied by an algebraic model.

But this method for establishing consistency only shows that if algebra is consistent, then is Hilbert's geometric system consistent. The proof is therefore only relative to the consistency of some other system and not an absolute proof.

139. What was the next step to avoid relative proofs ?

To avoid relative proofs of consistency Hilbert proposed a method which has come to be known as meta-mathematics. This method is well-equipped for investigating both consistency as well as completeness.

Hilbert and other mathematicians, therefore, hoped to develop each branch of mathematics in the axiomatic way in such a manner that it would be consisent as well as complete.

And the ultimate program was to develop one unified scheme for the whole of mathematics that was both consistent and complete.

This is known as "Hilbert's Program".

140. How far was Hilbert's Program successful ?

Meta-mathematical reasoning was successfully employed to establish consistency as well as completeness of more inclusive systems. For example, an absolute proof of consistency has been worked out for a system of arithmetic which permits addition of numbers, but not the multiplication.

Several such attempts were made to construct a proof also for the multiplication of numbers, but surprisingly all of them failed.

Finally in the year 1931, an Austrian mathematician Kurt Gödel showed that all such attempts must necessarily fail.

141. What did Gödel prove ?

or

What are the limitations of the axiomatic method ?

Gödel showed that the axiomatic method has certain inherent limitations regarding consistency and completeness.

He showed that consistency cannot be established within a system if whole of arithmetic is involved.

He also showed that the axiomatic method had another inherent limitation, that of incompleteness. Given any consistent set of arithmetical axioms, there are true arithmatical statements that cannot be derived from the set.

142. Any example to illustrate this ?

A simple example, viz., Goldbach's conjecture illustrates the point.

The conjuecture states that any even number (except 2, which is itself a prime) can be represented as the sum of two prime numbers.

Thus, 4 = 2 + 2, 6 = 3 + 3, 8 = 3 + 5,
10 = 5 + 5, 12 = 5 + 7, 14 = 7 + 7,
16 = 5 + 11, 18 = 5 + 13, 20 = 7 + 13,

Likewise, 50 = 19 + 31, 100 = 3 + 97, 200 = 3 + 197, etc.

Though no even number has been found which is not the sum of two prime numbers, yet no one has been able to find a proof valid for all even numbers.

The conjecture appears to be a true statement but cannot be derived from the axioms of arithmetic.

143. Will a different set of axioms not work ?

It may be suggested that the axioms be so modified or enlarged that this and other related theorems could be made derivable. But even if we added any finite number of airthmetical axioms, the enlarged system would still not suffice to yield all arithmetical truths.

There will always be further arithmetical truths which will not be derivable from the enlarged set. The axiomatic method is thus *essentially incomplete.*

Gödel further showed that for systems of the most important kind, consistency is incompatible with completeness. Such systems, if consistent, must necessarily be incomplete.

Also, that if a system is complete (for example, one which involves only addition and not multiplication of arithmetic), it can be shown to be inconsistent.

144. What is the essence of Gödel's discovery ?

The essence of Gödel's discovery is that no logical system which is both consistent and complete can be devised.

Before this discovery mathematicians had cherished the hope of evolving a consistent body of mathematics wholly included in one axiomatic setting.

The discovery meant a severe jolt to such a hope.

Thus Gödel did to logic in the year 1931 what Heisenberg* had done to Physics by his uncertainty principle just four years earlier in 1927.

145. What is the implication of the discovery ?

The implication is disturbing because the discovery has weakened the belief that mathematical truth is unerring and perfect.

This is because mathematical truth derives its force from an interplay of axioms called proofs, but when the axiomatic approach itself, which churns out such proofs, gets under scrutiny and suspicion, the picture obviously turns unreliable and dismal.

146. What is meant by Hilbert's formalism ?

Axiomatization of mathematical systems led to the view that mathematics is to be regarded as a mere game with mere marks on paper played according to certain definite rules, the game and the marks treated as devoid of any meaning or interpretation.

The systems are thus regarded as formalized in this sense and the outcome was Hilbert's formalism.

147. What is the advantage of this formalism ?

The advantage of regrading mathematical systems as formalized systems is that thereby one escapes a great many confusing and unnecessary questions, which otherwise are fundamental and cannot be summarily dismissed.

---

* Heisenberg's uncertainty principle implies that the very act of observing an elementary particle disturbs it in an unpredictable manner. This puts a limit to the powers of experimental inference.

148. What are such questions ?

To elucidate, consider the following questions:

What are numbers ?

Do numbers exist ?

How do we know that the laws of numbers are true ?

These and such others are important questions, but in a formalized system they become unnecessary and out of place.

The formulas of the system, then, mean anything, they are neither true nor false and make no claims regarding the existence of anything.

149. How did Gödel prove his result ?

Gödel numbered the symbols, formulas, and sequences of formulas, i.e., proofs in Hilbert's formalism in a certain way called Gödel numbering, and thus transformed all assertions into arithmetical propositions.

His method consisted of a set of rules for making a one-one correspondence between the integers and the various symbols or combinations of symbols. He was then able to show that the consistency of arithmetic is undecidable by any reasoning within the formalism of arithmetic.

To follow the actual proof, forty six preliminary definitions and several important lemmas have to be first mastered before one can proceed onwards.

The proof is difficult and the reasoning too complex to be followed by a non-mathematician.

150. Is Godel's work only negative in import ?

No.

Godel's work introduced a new technique of analysis in the foundations of mathemtics and gave rise to a very important branch of mathematics, namely, Proof Theory.

The technique has really aroused such vigorous activity in mathematical logic that its ultimate outcome is difficult to foretell.

Godel's work has actually spurred, rather than daunted mathematical creativity.

151. What is the moral of this great discovery ?

This monumental discovery has revealed the intrinsic limitations of the deductive method. It is widely regarded as one of the greatest intellectual accomplishments of the 20th century.

Nevertheless it need not cause dejection or disillusionment.

It only sigmfies that deeper and more sophisticated methods of investigation are yet to be devised, since creative reason admits of no limitations.

152. Has, then, the axiomatic method been abandoned ?

No, far from that. On the contrary, it is recognised as an accepted mode of exhibiting logical framework of any mathematical model.

In fact, Godel's result does not at all hamper our daily work, nor does it pose any threat to the bulk of mathematics which is in use everyday and almost everywhere.

153. What use is it adopting axiomatic method when a consistent system cannot be complete ?

It is true that for a considerable number of branches of mathematics we cannot get complete systems but the incomplete ones we can get are also highly enlightening. The advantages accruing therein are also exteremely fruitful.

The incomplcteness of the system does not in any way hamper its utility.

154. Why is the axiomatic method so widely used even when it suffers from inherent limitations ?

The method and its limitations are a part of the foundations of mathematics while its wide use is due to its extremely fruitful applications.

It is, thereore, always advisable to make distinction between mathematics and the applications of mathematics.

For example, a mathematical system which we call geometry is not necessarily a description of actual space. To assert that a particular type of geometry is a description of physical space is really a physical assertion, and not a mathematical statement.

Therefore in vast applications of mathematics one need not feel concerned about mathematical existence and like concepts, which truly belong to the domain of the foundations of mathematics.

155. What is it that is relevant for the applications of mathematics ?

What is relevant or important for applications is that the axioms and concepts of a mathematical system should match nicely with statements about actual objects and that it should be possible to verify the statements physically.

Godel's result is not in any way concerned with the applications of mathematics. It is the outcome of a deep investigation into the foundations of mathematics in general and mathematical existence in particular.

156. What precisely is meant by mathematical existence ?

We have seen that points and straight lines of geometry are abstractions of their physical counterparts and need not resemble them.

Likewise, it is not essential that mathematical entities need have close connection with objects of the physical world.

This brings out how mathematical existence differs from physical existence.

In the applications of mathemtics *if* the physical model fits the mathemtical model, mathematical results can be utilized, but complete correspondence between the two need not be taken for granted.

Applications are concerned with physical existence but mathematical models are concerned with mathematical existence only.

157. What set of axioms is sufficient for the algebra of schools ?

Algebra of schools deals mostly with numbers. The properties of numbers and the usual operations on them can be developed from the following set of axioms :

1. For any two numbers, their sum is uniquely determined.
2. For any two numbers, their product is uniquely determined.
3. There exists a number zero with the property $a + 0 = a$.
4. For every number a, there exists a number x such that $a + x = 0$.
5. Addition is commutative, i.e., $a + b = b + a$.
6. Addition is associative, i.e., $a + (b + c) = (a + b) + c$.

7. Multiplication is commutative, i.e., ab = ba.
8. Multiplication is associative, i.e., a (bc) = (ab) c.
9. Multiplication is distributive, i.e.,

$$a\,(b + c) = ab + ac,$$
$$(b + c)\,a = ba + ca.$$

10. For every a and every b not equal to zero, there exists a unique number x such that b x = a.

Any system of quantities which statisfies these ten conditions is called a *field.*

Examples of fields are the set of all rational numbers, the set of all real numbes, and the set of all complex numbers.

In each of these cases the numbers of the set when added and multiplied give a number which is a member of the same set, and the operations statisfy the ten conditions.

Apart from these examples there are many quantities of another nature which also form fields. Algebraic fractions, for example, also obey these ten conditions and so form a field.

158. How are new axiom systems formed ?

New axiom systems can be obtained by suppressing one or more axioms of a given system.

For example, by suppressing axiom 7, we get a system which is obeyed by the algebra of matrices, where the product of two matrices depends upon the order in which they are multiplied.

New axiom systems can also be obtained from a given system by changing one or more of its axioms in a suitable manner

The derivation of an axiom system for non-Euclidean geometry from axioms for the Euclidean geometry, by replacing the parallel axiom by one of its denials, is an example of obtaining new axiom systems in this manner.

❖❖❖

2

Algebra and Algebras

1. Geometry was developed in the axiomatic form but not Arithmetic and Algebra. Why?

The reason for this lies in their origins.

Geometry was discovered by the Egyptians as a result of their measurement of land. In the 7th century B.C. geometry passed from Egypt to Greece, where it gradually developed into a mathematical theory.

Thus geometry as a mathematical theory is of Greek origin. The Greeks attached great value to the proofs and therefore developed geometry along the axiomatic lines.

Our mathematics of numbers has its orgin in the mathematics of the Hindus, the Arabs and the Babylonians.

They did not concern themselves with giving proofs so that the mathematics of numbers came down to us merely in the form of a collection of rather unconnected rules of calculation.

The modern trend is to present all mathematics studies in the axiomatic form.

2. What is the meaning of the word 'arithmetic' ?

The word 'arithmetic' means the 'art of calculation' so that the study of arithmetic in our elementary schools is a collection of the solutions of various problems and the rules of calculation.

But with the passage of time arithmetic transformed into the theory of numbers.

3. Is arithmetic an abstraction ?

Arithmetic embodies the earliest attempts of the human mind at abstraction.

Thus when we say, $2 + 3 = 5$, it is a statement not about particular objects like pencils or coins, but for all countable objects maintaining their separate identity.

Here the nature of the objects, i.e., whether they are pencils or coins or plants or anything else, or living or non-living etc., loses its relevance, and the statement turns out to be true in a general way.

The numbers have been given names (one, two, three,....) and symbols (1, 2, 3,....) and are being used as concrete objects so persistently that we are apt to forget that we are dealing with concepts and not with concrete objects.

4. Is the statement 2 + 3 = 5 true for all kinds of objects ?

No. If the objects do not maintain their separate identify, the statement may not be true for them.

For example, 2 drops of water added to 3 drops of water may produce only one drop of water— a big drop.

Like wise, if 2 tigers and 3 rabbits are put together in a cage, 2 animals only will be visible in the cage after some time—the two tigers should have feasted on the helpless rabbits.

Again, a force equal to 2 units and another equal to 3 units, both applied to a body, may add up to anything between one unit or five units of force depending upon the angle between them

If they act in opposite directions, their sum will be one unit, but if they act in the same direction, their sum will be 5 units.

Their sum, however, will be 4 units if the angle between them is about 75 degrees and a half.

5. What is meant by the extension of the number concept ?

First came numbers connected with concrete objects so that the number concept was initially confined to whole numbers only. Fractions were naturally to follow and introduction of a symbol for zero was a great event, but negative numbers were admitted to the number–fold with great reluctance.

Such numbers are collectively known as rational numbers.

Once this relaxation began, it was the turn of the irrational and the complex numbers to be so recognized.

An irrational number is one not expressible as quotient of whole numbers.For example, $\sqrt{2}$ is an irrational number.

The rational and the irrational numbers are both known as real numbers.

A complex number is any number of the form $a + bi$, where a and b are real numbers, and i stands for the square root of minus one, so that $i^2 = -1$.

6. What are transcendental numbers ?

Irrational numbers which are not the roots of any algebraic equation are called transcendental numbers.

e and π are such numbers.

$e = 2.71828.....,$ $\pi = 3.14159.....,$

the dots at the end simply mean that the sequence does not end there but goes on indefinitely.

Regarding transcendental numbers there is an interesting result due to Gelfond who proved in 1934 that α^β is transcendental if α is algebraic and neither zero nor one, and β is algebraic and not rational.

Thus $2^{\sqrt{2}}, 3^{\sqrt{2}}, 5^{\sqrt{3}}$ are transcendental numbers. But if α and β are both transcendental it is not known whether α^β is transcendental. For example, it is not known whether e^e or π^π or π^e is trancendental.

However, $e^{i\pi} = -1$ is a result elegant par excellence.

7. Why is algebra called generalised arithmetic ?

An example will elucidate.

At school, a child learns that if 1 is subtracted from the square of a number, it is equal to the product of the numbers immediately preceding and succeeding the number.

Thus,

$$4^2 - 1 = (4 - 1)(4 + 1),$$
$$5^2 - 1 = (5 - 1)(5 + 1),$$
$$6^2 - 1 = (6 - 1)(6 + 1).$$

It is obvious that the proposition is true if in place of 4, 5 or 6, any other number is taken.

If a new symbol, say x, is introduced to represent any number, and not a number in particular, then the proposition can be written in a general way, thus

$$x^2 - 1 = (x - 1)(x + 1).$$

The introduction of the symbol x is the biginning of algebra.

8. Wherein lies the power of algebra ?

Algebra gains much of its power from dealing symbolically with elements, operations and relationships.

Symbols x, y, z, etc., are used as elements, addition and multiplication are chiefly employed as operations, and equality as the relationship normally connecting the elements.

Thus $x + x = 2x$, and $x + y = y + x$,

no matter what numbers x and y represent.

9. Has algebra also been generalized ?

The symbol x, used to represent any number, came to assume tremendous potential. To start with, it gave rise to algebraic equations, which dominated the scene for so long that until about a century and a half ago, algebra was just the theory of equations.

Later x was not restricted to numbers only but was employed to represent other entities also, and the operational signs for addition and multiplication allowed to acquire new meanings depending upon the kind of entity being considered.

The entity, therefore, determined the meaning to the attached to the signs + and x.

Vectors and matrices are two familiar examples of such entities. These will be referred to a little later.

This gives a generalized picture of what algebra originally stood for.

10. How is it different from the original form of algebra ?

In elementary algebra, the letters denote ordinary numbers, and the signs of operation, viz.+ and x, stand for ordinary addition and multiplication. But in the generalised form the letters denote any entity, and the signs of operation any rules of combination relevant to the entity.

11. What is abstract algebra ? Is it a further generalization ?

In abstract algebra, even these entities lose their meaning in the sense of being magnitudes and one speaks more generally of "elements" on which operations similar to the algebraic ones can be performed.

An example of such elements is provided by two motions applied one after the other, which together will be equivalent to certain single motion.

To illustrate, let R_1 denote the rotation of a square about its centre through 90°, R_2 that through 180°, and R_3 that through 270°, then the rotation R_1 Followed by R_2 will be equivalent to the single rotation R_3.

Another example is that of two algebraic transformations, that will produce the same result as a single transformation.

To illustrate, let the transformation T_1 denote translation, and let T_2 the rotation, then T_1 followed by T_2 will be equivalent to a single transformation T_3.

If, therefore, for a certain set of "objects", denoted by letters, certain operations can be defined in accordance with certain rules, then an algebraic system is said to be defined. Algebra has therefore come to be identified as the study of various algebraic systems, and is then known as abstract or axiomatic algebra.

12. Why is it called abstract or axiomatic ?

It is abstract because we are not concerned about what the letters in the algebraic system denote. What is important is that the axioms or laws are satisfied by the operations. And it is axiomatic because it is constructed solely from the laws or axioms stated in the beginning.

Two such algebraic systems are known as groups and rings.

The names seem a little strange in the beginning, but a little acquaintance wears off the initial reaction. We shall refer to them a little later.

13. Which fields of study use axiomatic algebra ?

Topology, Functional analysis, Quantum mechanics and contemporary physics are a few of the several important fields, where axiomatic algebra has proved to be the most powerful instrument for investigations.

14. Arithmetic as the theory of numbers ! What does the theory of numbers deal with ?

Elementary theory of numbers deals with the following :

Composite numbers and rules of divisibility, Prime numbers and their occurrence, The fundamental theorem of arithmetic, Fermat's Theorem, Wilson's Theorem, Fermat's Last Theorem,

Pythagorean numbers,

Properties of large numbers.

By numbers is implied, here, whole numbers or positive integers.

15. What are composite and prime numbers ?

Some numbers can be resolved into smaller factors,

e.g., 15 = 3 x 5, but not 11 or 17.

Numbers which can be resolved into smaller factors are called composite numbers, while those which cannot be so resolved are called prime.

16. What about the number 1 ? Is it a prime number ?

A prime number is one that has 1 and itself as divisors.

For example, the prime number 7 has two divisors, 1 and 7, though they are said to be the trivial divisors.

Therefore, it 1 were a prime number, it would have exactly two divisors If 1 were a composite number, it would have more than two divisors. But 1 has exactly one divisor, hence it is neither a prime nor a composite number.

17. What is meant by the rules of divisibility ?

The following are the rules of divisibility. One learns them at school.

1. A number is divisible by 2, if its unit's place is divisible by 2. Thus all numbers ending with 0,2,4,6 or 8 are divisible by 2, as in 530 and 138.
2. A number is divisible by 4, if the last two digits on the right are 00 or divisible by 4, as in 300 and 528.
3. A number is divisible by 8, if the last three digits on the right are 000 or divisible by 8, as in 3000 and 3240.
4. A number is divisible by 5, if the last digit on the right is 0 or 5, as in 240 and 235.
5. A number is divisible by 25, if the last two digits on the right are 00 or divisible by 25, as in 300 and 425.
6. A number is divisible by 3, if the sum of thc digits in the number is divisible by 3, as in 231.

Here $2 + 3 + 1 - 6$, which is divisible by 3, and so is 231.

The reason for this is easy to see :

$$\begin{aligned}231 &= 2 \times 100 + 3 \times 10 + 1\\ &= 2 \times (99 + 1) + 3\,(9 + 1) + 1\\ &= 2 \times 99 + 2 \times 1 + 3 \times 9 + 3 \times 1 + 1\\ &= 2 \times 99 + 2 + 3 \times 9 + 3 + 1\\ &= (2 \times 99 + 3 \times 9) + (2 + 3 + 1)\\ &= (\text{a multiple of } 9) + (\text{sum of the digits}).\end{aligned}$$

A number is, therefore, divisible by 3, if the sum of its digits is divisible by 3.

7. A number is divisible by 9 if the sum of the digits in the number is divisible by 9, as in 477.

   Here 4 + 7 +7 = 18, which is divisible by 9, and so is 477.

   The reason in this case is also similar to the one for the number 3.

8. A number is divisible by 11, if the difference between the sums of the digits in the odd and even places is 0 or a multiple of 11.

   Consider the number 1 8 3 9 5 5 2.

   Sum of the digits in the odd places = 1 + 3 + 5 + 2 = 11,

   Sum of the digits in the even places = 8 + 9 + 5 = 22,

   difference = 22 – 11 = 11, which is divisible by 11,

   ∴ 1 8 3 9 5 5 2 is divisible by 11.

18. Any other rules ?

Yes, the following are interesting :

1. The product of two numbers is equal to the product of their greatest common divisor and their least common multiple.

   Thus if 12 and 18 be the two numbers, their g. c. d and l. c. m are 6 and 36 respectively, and 12 x 18 = 6 x 36 = 216.

2. The product of two consecutive integers is divisible by 2, i. e, n (n + 1) is divisible by 2, where n is any integer.

3. The product of three consecutive integers, i.e., n (n + 1) (n + 2) is divisible by 2 x 3, i. e., 6.

4. The product of four consecutive integers, i. e,

   n (n + 1) (n + 2) (n + 3) is divisible by 2 x 3 x 4, i. e., 24.

5. The product of r consecutive integers is divisible by

   2 x 3 x 4 x...............x r or $\lfloor r$.

   The product 1. 2. 3.......r is called factorial r, and is denoted by the symbol $\lfloor r$ or r !

6. For all odd numbers n, the number $n^2 - 1$ is divisible by 8.

For, if n is odd, n – 1 must be even and divisible by 2. Also, n + 1 is the next even number, and therefore, divisible by 4. The product is, thus, divisible by 8.

19. How many primes are there ?

The prime numbers are infinite in number.

The primes less then 100 are, in order :

2, 3, 5, 7, 11, 13, 17, 19, 23, 29, 31, 37, 41, 43, 47, 53, 59, 61, 67, 71, 73, 79, 83, 89 and 97.

Some primes greater then 100 are :

101, 103, 107, 109,......., 211,....., 307,....., 401,......, 503,........, 601,......., 701,........, 809,........, 907,........, 65537,......, 510 511,.........

20. Is there the greatest prime ?

The question whether the above sequence has any end, or that whether the primes are infinite remained unanswered for quite some time, when Euclid showed that they must be infinite in number, and that there is no greatest prime.

21. How did Euclid show that the primes are infinite in number ?

The reasoning is as follows :

If there were only a finite number of primes, there must be the greatest of then all, say P, Then the number.

(2 x 3 x 5 x 7 x 11 x.....................x P) + 1

will leave the remainder 1 on being divided by each of 2, 3, 5, 7, 11, ..., P.

The above number is, therefore, not divisible by any of these primes. It is therefore a prime itself or is divisible by a prime greater than P. In either case, P is not the greatest prime. There are, therefore, an infinite number of primes.

22. What is the method of computing primes ?

The method of computing primes up to any number N is simple. First we write all the numbers from 1 to N, thus

1, 2, 3, 4,..................., N,

and cross out, first the number 1, then all the numbers except 2 which are multiples of 2, then all except 3 which are multiples of 3, then all except 5 which are multiples of 5, then all except 7 which are multiples of 7, and so on. The multiples of 4, 6, etc. have already been crossed out. The remaining numbers will then be primes.

23. How are primes distributed ?

Though primes are infinite, they become more and more rare as we encounter larger and larger numbers. But their distribution is extremely irregular, because whereas two consecutive primes may differ by 2 only, two consecutive primes may differ by a million also.

For, consider the numbers $\underline{|10} + 2$, $\underline{|10} + 3$, $\underline{|10} + 4, \ldots \underline{|10} + 10$, which are divisible by 2, 3, 4,....., 10 respectively. We may in this way construct as many consecutive composite numbers as we wish, even a million or more, none of which will be a prime. On the other hand, the prime numbers 1, 000, 000, 009, 649 and 1, 000, 000, 009, 651 are so close that they differ by 2 only.

24. How many primes are there between any number and its double ?

Between any number greater than 1 and its double there is always at least one prime number.

Joseph Bertrand had conjectured this result and verified it empirically from the tables up to quite large values of n, but was actually proved by Chebychev.

25. How many primes are there less than a given number ?

An estimate of the number of primes less than a given number has also been made.

The primes less than 20 are 2, 3, 5, 7, 11, 13, 17, 19, i. e., 8 in number, so we say $\pi(20) = 8$.

Similarly, $\pi(100) = 25$, $\pi(200) = 46$, $\pi(300) = 62$, $\pi(400) = 78$, $\pi(500) = 95$, $\pi(600) = 109$, $\pi(700) = 125$, $\pi(800) = 139$, $\pi(900) = 154$, $\pi(1000) = 168$.

The list can be continued indefinitely, but it is not possible to give a simple formula for $\pi(x)$, where $\pi(x)$ denotes the number of primes less than x.

26. What is the prime number theorem ?

The prime number theorem states that for large x, the number of prime numbers less than x is about equal to $\frac{x}{\log x}$, where the logarithm is the natural logarithm.

The theorem was conjectured by Gauss in 1793, but proved by Hadamard and de la Valle'e Poussin a century later in 1896.

27. Is there any formula giving all the primes ?

No. Considerable effort has been made on inventing a formula which will give all the primes, but to no success.

Some instances may be recalled.

The expression n^2+n+17 is prime for all values of n from 1 to 16,

$2n^2+29$ is prime for values of n from 1 to 28,

n^2-n+41 is prime for values of n from 1 to 40,

and $n^2-79n+1601$ or $(n-40)^2+(n-40)+41$ is prime for values of n from 1 to 79.

Dirichlet proved that every arithmetical progression

$an+b$, $n = 0, 1, 2, 3,$.........,

where a, b are positive integers having no common factor greater than 1, contains an infinite number of primes.

For example there are an infinity of primes of the form $6n+1$, though, of course, not all of them primes. For $n=4$, $6n+1$ equals 25, which is not a prime.

It has however, been shown that no rational algebraical formula can represent prime numbers only.

28. Are all prime numbers alike ?

There are two forms of the primes.

All prime numbers except 2 are either of the form $4n-1$ or $4n+1$.

Of these, every prime of the form $4n+1$ can be expressed as the sum of two squares *in one way* only. Thus, $5 = 1^2+2^2$, $13 = 2^2+3^2$, $17 = 1^2+4^2$, $29 = 2^2+5^2$, $953 = 13^2+28^2$.

However, if a number of the form $4n+1$ can be expressed as the sum of two squares in two different ways, it cannot be a prime. For example, $545 = 17^2+16^2 = 23^2+4^2$, and 545 is not a prime.

No integer of the form 4n – 1 can be the sum of two squares, for example, 11 or 23 cannot be so expressed.

29. What is meant by the unsettled questions regarding primes?

Two simple – looking questions related to primes that have not yet been settled are the following :

One is whether or not there are an infinity of primes of the form n^2+1, where n is a whole number.

If we give to n successively the values 1, 2, 3, 4, 5, 6, 7, 8, 9, 10, etc., $(n^2 + 1)$ takes up the values 2, 5, 10, 17, 26, 37, 50, 65, 82, 101, etc. Out of these some are prime numbers while others are not. The question is whether a stage will be reached when this process will stop yielding prime numbers.

The other is the conjecture of Goldbach asserting that every even number greater than 2 is a sum of two primes, for example, 40 = 11+29. The conjecture is confirmed by the tables as far as they go, but has never been proved.

30. The Fundamental Theorem of Arithmetic ! What is it ?

A property shared by every integer greater than 1 is that it is either a prime itself or can be factored into primes in only one way.

The result that for every positive integer this factorization is unique is known as the Fundamental Theorem of Arithmetic.

Thus, 30 can be factored as 2 × 3 × 5 only and in no other way, a different order of the factors, say 3 × 5 × 2, is not regarded as a different factorization.

The theorem is also known as the Unique Factorization Theorem.

31. What are twin primes ?

An interesting phenomenon is the occurrence of prime pairs also known as twin primes.

A prime pair is a pair of primes whose difference is 2, such as 11 and 13.

The prime pairs, less than 1000, are, in order :

(3, 5), (5, 7), (11, 13), (17, 19), (29, 31), (41, 43) (59, 61) (71, 73), (101, 103), (107, 109), (137, 139), (149, 151), (179, 181), (191, 193), (197, 199), (227, 229), (239, 241), (269, 271), (281, 283), (311, 313),

(347, 349), (419, 421), (431, 433), (461, 463), (521, 523), (569, 571), (599, 601), (617, 619), (641, 643), (659, 661), (809, 811), (821, 823), (827, 829), (857, 859), and (881, 883).

32. Are the twin primes also infinite in number ?

The thirty five pairs given above are those that lie between 1 and 1000. But the list can be continued indefinitely.

Another such pair is (4049, 4051).

Still another pair is (1, 000, 000, 009, 649 and 1, 000, 000, 009, 651).

It has been conjectured that the number of these is infinite, but no proof has been found.

33. What property is common to the twin primes ?

All prime pairs, with the exception of the first prime pair, i.e., (3, 5), possess the remarkable property that the sum of the primes in the pair is always divisible by 12.

For example, (5, 7) has the sum 12, (11, 13) has 24, (17, 19) has 36, and so on, each sum being divisible by 12.

34. How many divisors has a composite number ?

Let $N = a^p b^q c^r$ be a composite number, where a, b, c are different prime numbers and p, q, r are positive integers.

The number of divisors is, then, $(p + 1)(q + 1)(r + 1)$.

How ?

Consider the product :

$$(1 + a + a^2 + \ldots\ldots + a^p)(1 + b + b^2 + \ldots\ldots + b^q)(1 + c + c^2 + \ldots\ldots + c^r).$$

The total number of terms in this product is $(p + 1)(q + 1)(r + 1)$ and each term of the product is a divisor of the given number. Hence the number of divisors is $(p + 1)(q + 1)(r + 1)$.

Also, no other number can be a divisor.

This also includes as divisors, both 1 and the number N itself.

35. How is this property generalized ?

If $N = a^p b^q c^r d^s \ldots$, the number of divisors will similarly be equal to $(p + 1)(q + 1)(r + 1)(s + 1)\ldots$, the divisors 1 and N both included.

36. How many divisors has the number 30, and what are they ?

Since $30 = 2 \times 3 \times 5 = 2^1 \times 3^1 \times 5^1$,

the number of divisors $= 2 \times 2 \times 2 = 8$.

The divisors are 2, 3, 5, 6, 10, 15 ; 1 and 30.

37. How many divisors has the number 7056 ?

Since $7056 = 2^4 \times 3^2 \times 7^2$,

The number of divisors $= (4 + 1)(2 + 1)(2 + 1) = 5 \times 3 \times 3 = 45$.

Excluding the trivial divisors 1 and 7056,

the number of proper divisors = 43.

38. How is the highest power of a prime number contained in $\underline{|n}$ determined ?

An example will illustrate the method.

Let us find the highest power of 3 in $\underline{|100}$, i.e., in the product 1.2.3........ 100.

The integer 3 occurs only in the integers 3, 6, 9,....., 99, i.e., every third integer.

Their number is, therefore, given by the quotient of 100 divided by 3, i.e., 33.

3 occurs a second time in the integers 9, 18, 27,....., 99. their number is the quotient of 100 divided by 9, i.e., 11.

3 occurs a third time in the integer 27, 54, 81.

Their number is the quotient of 100 divided by 27, i.e., 3.

3 occurs as a factor four times in 81 only.

Hence the highest power required = 33 + 11 + 3 +1 = 48.

Thus to find the highest power of a prime p contained in $\underline{|n}$, we find the quotients of n divided in turn by p, p^2, p^3,....., and add them all.

Likewise, it can be found that the highest power of 7 contained in $\underline{|1000}$ is given by 164.

39. What is Fermat's Theorem ?

If p is a prime number, and N be prime to p, then $N^{p-1} - 1$ is a multiple of p.

This is known as Fermat's Theorem.

Since N is prime to p, the above expression may be multiplied by N, and we get the following result :

$N^p - N$ is divisible by p for every prime p.

Thus $n^2 - n$ is divisible by 2.

In words, this means that the difference between the square of a number and the number itself is always an even number.

Similarly, $n^3 - n$, $n^5 - n$, $n^7 - n$, $n^{11} - n$, etc. are divisible by 3, 5, 7, 11, etc. respectively, but similar results do not hold good for $n^4 - n$, $n^6 - n$, etc., as 4, 6.etc., are not prime numbers.

40. But how is $n^5 - n$ divisible by 30, and not by 5 only ?

Since 5 is a prime number, therefore, $n^5 - n$, is divisible by 5 by Fermat's Theorem.

Also,

$$\begin{aligned} n^5 - n &= n\,(n^4 - 1) \\ &= n\,(n^2 - 1)\,(n^2 + 1) \\ &= n\,(n-1)\,(n+1)\,(n^2 - 1) \\ &= (n-1)\,n\,(n+1)\,(n^2+1). \end{aligned}$$

Now, $(n-1)\,n\,(n+1)$ denotes the product of three consecutive natural numbers, and is divisible by $\underline{|3}$ or 6.therefore, $n^5 - n$ is divisible by 5×6, i.e., 30.

For the same reason, $n^7 - n$ is also divisible by 7×6 i. e. 42, and not by 7 only.

41. What other results follow from Fermat's Theorem ?

The following are the results :

1. Every square number is of the form 5n or $5n \pm 1$, where n is a positive integer.
2. Every number which is a perfect cube is of the form 9n or $9n \pm 1$.
3. A number which is both a perfect square and a perfect cube is of the form 7 n or 7 n+1.

42. What is Wilson's Theorem ?

It states that :

The number $\underline{|n-1} + 1$ is divisible by n, if, and only if, n is prime.

Thus, for $n = 5$, $\lfloor\underline{n-1} + 1$ equals 25, which is divisible by 5, since 5 is prime.

But, if $n = 6$, $\lfloor\underline{n-1} + 1$ equals 121, which is not divisible by 6, since 6 is not prime.

43. How is mathematical induction used to prove divisibility ?

The method of mathematical induction where we pass from particular to general statement can sometimes be used to prove some results in divisibility.

As an illustration, we prove that $3^{2n}-2n-1$ is divisible by 2 only for all positive integral values of n.

Let us denote the expression by f (n), then

$$f(n) = 3^{2n} - 2n - 1 \qquad \text{--- (1)}$$

Changing n into (n + 1), we have

$$f(n+1) = 3^{2n+2} - 2(n+1) - 1$$
$$= 9.\,3^{2n} - 2n - 3 \qquad \text{---(2)}$$

multiplying (1) by 9 and subtracting from (2), we get

$$f(n+1) - 9f(n) = -2n - 3 - 9(-2n - 1)$$
$$= -2n - 3 + 18n + 9$$
$$= 16n + 6$$
$$= 2(8n + 3),$$

Thereore, if f (n) is divisible by 2, so also is f (n + 1)

But f (1), i.e., $3^2 - 2 - 1 = 6$, is divisible by 2,

∴ f (2) is also divisible by 2, and so is f (3), and so on.

The result is thus true universally.

The following results can be similarly proved:

(i) $10^n + 3.\,4^{n+2} + 5$ is divisible by 9,

(ii) $3^{4n+2} + 5^{2n+1}$ is divisible by 14,

(iii) $3^{2n+2} - 8n - 9$ is divisible by 64,

(iv) $3^{2n+5} + 160n^2 - 56n - 243$ is divisible by 512,

(v) $5^{2n+2} - 24n - 25$ is divisible by 576.

44. What are Pythagorean numbers ?

Positive integers x, y, z are called Pythagorean numbers if they satisfy the equation : $x^2 + y^2 = z^2$.

Two familiar examples of such numbers are 3, 4, 5, and 5, 12, 13. Here we have, $3^2 + 4^2 = 5^2$; and $5^2 + 12^2 = 13^2$.

Pythagorean numbers always form the sides of a right – angled triangle.

The most well – known property of a right – angled triangle is given by the Theorem of Pythagoras. The theorem states that the sum of the squares on the sides containing the right angle is equal to the square on the hypotenuse.

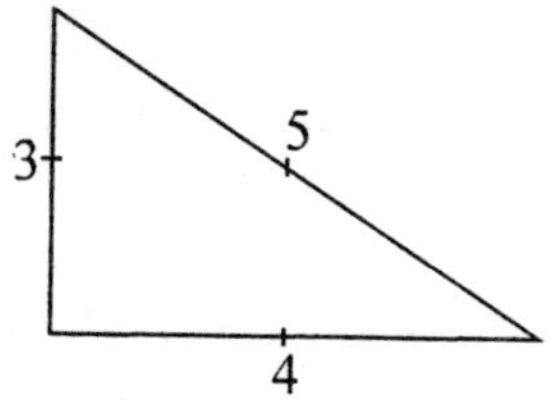

By Pythagoras Theorem : $3^2 + 4^2 = 5^2$.

All such numbers are given by

$$x = m^2 - n^2,$$
$$y = 2mn,$$
$$z = m^2 + n^2,$$

Where m, n are any positive integers, and m greater than n.

45. What about the sum of higher powers of integers ? or What is Fermat's Last Theorem ?

It is natural next to search for positive integers x, y, z such that

$x^3 + y^3 = z^3$,

$x^4 + y^4 = z^4$,

$x^5 + y^5 = z^5$, and so on.

All these cases are included in the following :

To find integers x, y, z such that $x^n + y^n = z^n$, where n is an integer greater than 2.

Fermat in about 1637 devoted himself to this problem and came to the conclusion that it is impossible to find such integers.

This result is known is Fermat's Last Theorem.

But he mentioned that he had found an admirable proof of this, and that the margin of the book where he wrote was too narrow to contain it. Fermat was accustomed to record some of his ideas in the margin of his mathematics book.

46. Has the proof since been rediscovered ?

Several brilliant mathematicians have since worked for more than three hundred years to discover the proof but without success.

The theorem has been proved for several values of n, and no exception to the theorem has ever been found, but a general proof true for all values of n has so far evaded all attempts.

47. Can every positive integer be expressed as the sum of four squares ?

An interesting property true for all positive integers is that any integer can be expressed in the form $x^2 + y^2 + z^2 + u^2$, zero values of x, y, z, u being admissible.

Thus

$$\begin{aligned} 1 &= 0^2 + 0^2 + 0^2 + 1^2, \\ 2 &= 0^2 + 0^2 + 1^2 + 1^2, \\ 3 &= 0^2 + 1^2 + 1^2 + 1^2, \\ 4 &= 1^2 + 1^2 + 1^2 + 1^2, \\ 5 &= 0^2 + 0^2 + 1^2 + 2^2, \\ 6 &= 0^2 + 1^2 + 1^2 + 2^2, \\ 7 &= 1^2 + 1^2 + 1^2 + 2^2, \end{aligned}$$

etc.

Also

$$\begin{aligned} 50 &= 0^2 + 0^2 + 1^2 + 7^2, \\ 234 &= 2^2 + 5^2 + 6^2 + 13^2, \\ 2011 &= 13^2 + 16^2 + 19^2 + 35^2, \end{aligned}$$

etc.

48. Is the representation unique ?

No. It is possible to express a number in such form in more than one way, thus

$$\begin{aligned} 10007 &= 99^2 + 14^2 + 3^2 + 1^2, \\ &= 74^2 + 65^2 + 15^2 + 9^2, \\ &= 62^2 + 59^2 + 51^2 + 9^2, \end{aligned}$$

49. Do such results exist for third and higher powers of the integers?

Investigations have been made in this direction since 1770 and the results continually improved.

The best results achieved so for state that every sufficiently large integer N is a sum of 9 cubes, 19 fourth powers, 41 fifth powers, 87 sixth powers, 193 seventh powers, 425 eighth powers, 949 ninth powers or 2113 tenth powers.

The limit beyond which N must lie has not been determined, but it must be unusually large.

50. What is Goldbach's conjecture regarding large numbers ?

In 1742 Goldbach conjectured that every sufficiently large odd number N is representable as the sum of three primes, i. e.,

$$\text{odd } N = p_1 + p_2 + p_3,$$

but was actually proved by Vinogradov in 1937.

If we add 3 to both the sides in this relation, we have

$$\text{even } N = p_1 + p_2 + p_3 + 3,$$

i.e., every sufficiently large even number is representable as the sum of four primes.

It is also known that every sufficiently large integer is the sum of at most 20 primes.

51. Algebra as the Theory of Equations ! What is meant by solving an equation ?

Consider the following problems :

1. A is twice as old as B. Ten years ago he was four times as old as B. What are their present ages ?
2. A had twice as much money as B. After each spent Rs. 10 on purchases, A found that he had now four times as much as B. How much did each have initially ?
3. A goes twice as far as B. If each had covered 10 miles less, A would have gone four times as far as B. How far did each go ?

The unknown entity like age, money, distance in these problems is given a letter name x and the problem stated using the symbol x.

This statement about x usually joins two expressions by an equality sign, hence it is called an equation. This equation is true for certain value or values of the unknown x, and not for other values.

Solving the equation means to determine those values of x for which the equation is true. For example, the equation $4x = 12$ is true for $x = 3$

only, so that 3 is called the solution or the root of the equation $4x = 12$.

52. How are these problems solved ?

In problem 1,

We suppose the age of B to be x years.

Then the age of A is 2x years.

Ten years ago, the age of A must be $(2x - 10)$ years.

And the age of B $= (x - 10)$ years.

According to the problem, age of A was 4 times the age of B,

$\therefore$ $(2x - 10) = 4(x - 10)$ or $2x = 30$ or $x = 15$.

$\therefore$ B's age is 15 years,

and A's age is double of B's, $\therefore$ 30 years.

In problem 2, we suppose that B had rupees x, then A's money was Rs. 2x.

Having spent Rs. 10 each, A is left with Rs. $(2x - 10)$, and B with Rs. $(x - 10)$.

According to the problem, A's money is now 4 times that of B,

$\therefore$ $2x - 10 = 4(x - 10)$ or $2x = 30$ or $x = 15$.

$\therefore$ B's money is rupees 15, and A's is double of B's i.e., Rs. 30.

In problem 3,

We suppose that B goes x miles, t hen A goes 2x miles.

If each had gone 10 miles less than what he did,

A should have gone $(2x - 10)$ miles, and B equal to $(x - 10)$ miles.

According to the problem, A's distance is equal to 4 times B's distance,

$\therefore$ $(2x - 10) = 4(x - 10)$ or $2x = 30$ or $x = 15$.

Thus B goes 15 miles and A goes 30 miles,

53. What is meant by the equation as a mathematical model ?

If may be observed that all the three problems discussed above have apparently different themes like age, money and distance travelled, but the same equation, viz., $2x - 10 = 4(x - 10)$ represents the apparatus needed to solve all of them.

The equation is thus a mathematical model which has *so much in common* with the problem that its solution is also the solution of the problem. Thus while we solve only the model, the problem gets solved.

54. What is meant by the model having "so much in common" with the problem ? Does the model not represent the problem completely ?

The set of natural numbers 1, 2,3,.... is the simplest example of a mathematical model. It is used for counting objects when all the properties of the objects are disregarded except their number.

But if other considerations are taken into account, they might lead to strange or unexpected conclusions as is clear from the following anecdote :

Under adult education program in a village, a school teacher was once trying to explain subtraction as follows :

Teacher : Out of 11 sheep if 7 jump over the fence, how many will remain behind ?

Student : None at all !

Teacher : Why ? If 7 of them jump over to the other side, 4 will still be on this side of the fence. Why not ?

To which the troubled student protested :

Student : Ma'am, you may be knowing mathematics, but you do not know sheep.

55. What is the procedure for solving problems in Algebra ?

To solve problems, they are turned into equations. How to solve equations is the central topic of Algebra, a brief account of which will now follow.

56. What is an equation of the first degree ?

An equation of the form $ax + b = 0$, where x is an unknown entity, is called an equation of the first degree.

It can be solved without any difficulty.

Thus, if $ax + b = 0$, then $ax = -b$, and $x = -\frac{b}{a}$.

In the problems discussed above, the equation $2x = 30$ is an equation of the first degree.

57. What is a quadratic equation ?

An equation of the second degree is called a quadratic. A quadratic in one unknown is of the form.

$$ax^2 + bx + c = 0.$$

It has two solutions, though they may sometimes coincide.

58. How is it solved ?

The principal device for solving a quadratic equation is a formula, which is derived as follows :

Each term of the equation is first divided by a. The third term $\frac{c}{a}$ is transferred to the other side with opposite sign and then $\frac{b^2}{4a^2}$ is added to both the sides, and square root taken of both the sides, thus :

$$ax^2 + bx + c = 0$$

or $x^2+\frac{b}{a}x+\frac{c}{a}=0$, on dividing both sides by a.

or $x^2 + \frac{b}{a}x = -\frac{c}{a}$

or $x^2 + \frac{b}{a}x+\frac{b^2}{4a^2} = \frac{b^2}{4a^2} - \frac{c}{a}$

or $\left(x+\frac{b}{2a}\right)^2 = \frac{b^2-4ac}{4a^2}$

$\therefore\ x + \frac{b}{2a} = \pm\frac{\sqrt{b^2-4ac}}{2a}$,

so that $x = \frac{-b\pm\sqrt{b^2-4ac}}{2a}$

The two solutions are $\frac{-b+\sqrt{b^2-4ac}}{2a}$ and $\frac{-b-\sqrt{b^2-4ac}}{2a}$

Example : $2x^2 - 7x + 6 = 0$

Here $a = 2$, $b = -7$, $c = 6$,

$\therefore$ the solutions are $\frac{7+\sqrt{49-48}}{4}$ and $\frac{7-\sqrt{49-48}}{4}$

or $\frac{7+1}{4}$ and $\frac{7-1}{4}$, i. e. 2 and $\frac{3}{2}$.

59. When were these methods of solution developed ?

Equations of the first degree are believed to have been solved by the Egyptians some 4,000 years ago, The quadratic equation was solved by the Hindus in early antiquity, but general equations of the third and the fourth degree were solved several centuries later by Italian algebraists of the 16 th century.

60. How many roots does an equation have ?

An equation of the first degree has one root, a quadratic two roots, a cubic three roots, and so on according to the degree of the equation.

In the early stage of the development of mathematics, only positive roots of equations were recognized and the negative roots were treated as false.

61. Do all algebraic equations have real roots ?

No. There are equations like $x^2 + 1 = 0$ which have no real roots.

The equation $x^2 + 1 = 0$ has two roots i and – i, where i denotes $\sqrt{-1}$, i. e. the square root of – 1.

To be able to say that every quadratic equation has two roots, it became necessary to recognize complex numbers as numbers in their own right – a status so far being continually denied to them.

A number of the form a + it is called a complex number. If a = 0, the number is sometimes called an imaginary number.

But the equation $x^2 - 2 = 0$ has two real roots $\sqrt{2}$ and $-\sqrt{2}$.

Such roots also baffled mathematicians till surds were admitted to the family of numbers.

62. What was gained by the extended number system ?

or

What is the Fundamental Theorem of Algebra ?

With the extended number system including positive whole numbers, fractions, negative numbers, surds and complex numbers, it became possible to enunciate a very important and elegant proposition known as the Fundamental Theorem of Algebra.

It states that any n th degree *algebraic* equation with real or complex coefficients always has at least one real or complex root.

It is called the Fundamental Theorem of Algebra because when it was first proved in 1799 by Gauss, the study of algebra was confined to the theory of equations only. Though the theorem is extremely important, the name is no longer justified in view of the vast change in the nature and scope of algebra.

A very useful consequence of this theorem is that every algebraic equation of the n th degree has not only one but exactly n roots. It is of course assumed that a repeated root is counted as many times as it is repeated.

63. Why is the Fundamental Theorem of Algebra called an existence theorem ?

It is called an existence theorem because it simply tells us the number of roots that exist for a given equation, but does not concern itself with a method of determining them.

64. Is the theorem true for all types of equations ?

No. The theorem is true only for algebraic equations since there exist some non-algebraical equations which have no roots at all !

For example, the equation $a^x = 0$, where a is real, has no root at all !

65. Which equations are called non-algebraic ?

The following are a few non-algebraic equations :

(i) $x + \log_{10} x = 5$,

(ii) $e^x - 3x = 0$,

(iii) $x^2 + 4 \sin x = 0$.

These equations are not algebraic as they contain logarithmic, exponential or trigonometrical expressions.

66. Has the number system been further generalized beyond the complex numbers ?

Attempts were made to further generalize the number concept but not with much success.

Quaternions and hypercomplex numbers were invented to achieve further generalization.

67. What is a quaternion ?

A quaternion is a symbol of the type a + bi + cj + dk, where a, b, c, d are real numbers, and i, j, k operational symbols.

Sum of two quaternions is easily defined. For example, the sum of two quatermions.

$x = x_0 + x_1 i + x_2 j + x_3 k$,

and $y = y_0 + y_1 i + y_2 j + y_3 k$,

is $x + y = (x_0 + y_0) + (x_1 + y_1) i + (x_2 + y_2) j + (x_3 + y_3) k$.

Product of two quaternions is defined by the use of the distributive law and the following conventions :

$$i^2 = j^2 = k^2 = -1,$$
$$ij = -ji = k,$$
$$jk = -kj = i,$$
$$ki = -ik = j.$$

They were invented by William R. Hamilton.

68. What is a hypercomplex number ?

A hypercomplex number is denoted by the expression

$$E_1 x_1 + E_2 x_2 + \ldots\ldots\ldots\ldots + E_n x_n,$$

where $x_1, x_2, \ldots\ldots\ldots\ldots, x_n$ are real numbers, and $E_1, E_2, \ldots\ldots, E_n$ operational symbols.

It is also called n-dimensional vector, and was created by Grassmann, a contemporary of Hamilton.

The theory of hypercomplex numbers includes that of quaternions, so that quaternions may be looked upon as a special case of hypercomplex numbers.

69. Why were these extensions of the number system not popular ?

Reasons were many.

The physicists and the applied mathematicians found them too general and too complex for their routine requirements.

Secondly, a much simpler mathematical instrument called Vector Analysis got evolved, which, because of its great power, came to be extensively applied in nearly every branch of mathematical physics and elsewhere.

Thirdly, the conventions used by Hamilton to define the product of two quatermions or the rules set up by Grassmann to combine two hypercomplex numbers did not enjoy the force of mathematical validity.

70. What then is the answer to the question : Can the number concept be further extended beyond the complex number system ?

The answer is No, and is a great landmark.

It was proved by Weierstrass about 1860, and more simply by Hilbert later, that *no further generalization in this particular direction is possible.*

We have reached the end of the road.

71. What is a cubic equation and how is it solved ?

An equation of the third degree is called a cubic.

The general cubic equation is

$$x^3 + ax^2 + bx + c = 0 \qquad \text{.....(1)}$$

It is first reduced to a cubic equation of the form

$$y^3 + py + q = 0, \qquad \text{.....(2)}$$

which does not contain the term with the square of the unknown.

This is easily done by setting $x = y - \frac{a}{3}$.

Now suppose $y = u + v$,

then cubing both the sides :

$$y^3 = u^3 + v^3 + 3uv(u + v)$$

or

$$y^3 = u^3 + v^3 + 3uvy,$$

putting y for u + v,

$$\text{or} \quad y^3 - 3uvy - (u^3 + v^3) = 0 \qquad \text{.....(3)}$$

Comparing equations (2) and (3), we have:

$$uv = -\frac{p}{3}, \text{ and } u^3 + v^3 = -q\,;$$

or

$$u^3 + v^3 = -q, \text{ and } u^3 v^3 = -\frac{p^3}{27}$$

Then u^3 and v^3 will be the roots of the quadratic equation,

$$t^2 + qt - \frac{p^3}{27} = 0 \qquad \text{.....(4)}$$

Solving this equation,

$$u^3 = -\frac{q}{2} + \sqrt{\frac{q^2}{4} + \frac{p^3}{27}}, \text{ and } v^3 = -\frac{q}{2} - \sqrt{\frac{q^2}{4} + \frac{p^3}{27}},$$

so that

$$y = u + v$$

$$\text{or} \quad y = \left\{-\frac{q}{2} + \sqrt{\frac{q^2}{4} + \frac{p^3}{27}}\right\}^{1/3} + \left\{-\frac{q}{2} - \sqrt{\frac{q^2}{4} + \frac{p^3}{27}}\right\}^{1/3}.$$

Since $x = y - \frac{a}{3}$, x is obtained by subtracting $\frac{a}{3}$ from the above.

72. Who evolved this remarkable method ?

This method of solving a cubic is generally known as Cardan's method.

Cardan obtained it from another mathematician Tartaglia under promise of secrecy but published it as his own in his book in 1545. Both Cardan and Tartaglia were Italians.

73. How is a cubic with numerical coefficients solved ?

Let us try to solve the following equation :

$$x^3 + 6x^2 + 9x + 4 = 0$$

First, let $x = y - 2$, then the given equation reduces to

$$(y - 2)^3 + 6(y - 2)^2 + 9(y - 2) + 4 = 0$$

or $y^3 - 3y + 2 = 0$, on simplification.

Now, let $y = u + v$, then $y^3 - 3uvy - (u^3 + v^3) = 0$, on cubing both the sides.

$\therefore$ $uv = 1$, and $u^3 + v^3 = -2$, on comparing with $y^3 - 3y + 2 = 0$

or $u^3 + v^3 = -2$, and $u^3 v^3 = 1$.

u^3 and v^3 are, therefore, the roots of the equation :

$t^2 -$ (sum of the roots) $t +$ product of the roots $= 0$,

or $t^2 + 2t + 1 = 0$,

or $(t + 1)^2 = 0$, giving $t = -1, -1$

$\therefore$ $u^3 = -1$, and $v^3 = -1$; giving $u = -1$ and $v = -1$.

$\therefore$ $y = u + v = -2$

Since $y = -2$ is a solution of the equation $y^3 - 3y + 2 = 0$,

$\therefore$ $y + 2$ must be a factor of this equation.

Dividing it by $y + 2$, we get the quadratic

$y^2 - 2y + 1 = 0$, or $(y - 1)^2 = 0$, giving $y = 1, 1$.

$\therefore$ $y = -2, 1, 1$.

Since $x = y - 2$, we have finally, $x = -4, -1, -1$.

74. Does this method always yield the solution ?

In case of cubics with numerical coefficients, this method gives the solution only when the cubic has either two imaginary or two equal roots, and fails to give the solution of a cubic all of whose roots are real and unequal.

A cubic of the later type is solved by the use of Trigonometry or by approximation methods.

75. How is an equation of the fourth degree solved ?

An equation of the fourth degree is also called biquadratic.

An example will best illustrate the method of solution.

Consider the equation :

$$x^4 - 10x^3 + 35x^2 - 50x + 24 = 0$$

The solution consists of first expressing the left hand side as difference of two squares and then as product of two quadratics.

we have $\quad x^4 - 10x^3 + 35x^2 - 50x + 24 = 0$

or $\quad x^4 - 10x^3 = -35x^2 + 50x - 24$

or $\quad x^4 - 10x^3 + 25x^2 = -10x^2 + 50x - 24$

or $\quad (x^2 - 5x)^2 = -10x^2 + 50x - 24.$

Introducing λ on the left and balancing by the right hand side, we get,

$$(x^2 - 5x + \lambda)^2 = (-10x^2 + 50x - 24) + \lambda^2 + 2\lambda(x^2 - 5x)$$

or $\quad (x^2 - 5x + \lambda)^2 = (2\lambda - 10)x^2 + (50 - 10\lambda)x + \lambda^2 - 24.$

Applying the condition that the terms on the right form a perfect square, we get what is called the auxiliary cubic in λ, from which λ can be determined.

Here, λ can be more easily determined by the consideration that since the terms on the right form a perfect square, $(2x - 10)$ and $(\lambda^2 - 24)$ must each be a perfect square.

We, therefore, put $(2x - 10)$ equal to 1, 4, 9, 16,.... etc. by turn and see which value of λ makes $\lambda^2 - 24$ also a perfect square.

Putting $2x - 10 = 4$ or $\lambda = 7$ makes $\lambda^2 - 24$ equal to 25, which is a perfect square, and we get

$$(x^2 - 5x + 7)^2 = 4x^2 - 20x + 25,$$

or $\quad (x^2 - 5x + 7)^2 = (2x - 5)^2,$

or $\quad x^2 - 5x + 7 = \pm(2x - 5),$

so that $\quad x^2 - 5x + 7 = 2x - 5$, and $x^2 - 5x + 7 = -2x + 5,$

i.e., $\quad x^2 - 7x + 12 = 0$, and $x^2 - 3x + 2 = 0.$

These are both quadratic equations, which on solving give $x = 3, 4$ and $x = 1, 2$ respectively.

The roots of the given biquadratic are thus 1, 2, 3, and 4.

76. Who evolved this method and when ?

This method of solving a biquadratic was given in 1540 by Ferrari, another Italian mathematician and a pupil of Cardan.

77. What is Des Cartes' method of solving a biquadratic ?

In 1637, Des cartes gave a method which was different from Ferrari's. He solved the equation by expressing it as a product of two quadratic expressions.

This method is applicable when either the term containing x^3 is absent, or has been removed by suitable substitution.

The following example illustrates the method.

Consider the equation :

$x^4 - 2x^2 + 8x - 3 = 0.$

Assume $x^4 - 2x^2 + 8x - 3 = (x^2 + kx + l)(x^2 - kx + m)$,
then simplifying and equating like coefficients,

we have,

$$\left.\begin{aligned} m + l - k^2 &= -2, \\ k(m - l) &= 8, \\ ml &= -3. \end{aligned}\right\}$$

i.e.,

$$\left.\begin{aligned} m + l &= k^2 - 2, \\ m - l &= \frac{8}{k}, \\ ml &= -3. \end{aligned}\right\}$$

Using the identity

$$(m + l)^2 - (m - l)^2 = 4\,m\,l$$

to eliminate m, l from these equations, we have

$$(k^2 - 2)^2 - \frac{64}{k^2} = -12.$$

Simplifying, $k^6 - 4k^2 + 16k^2 - 64 = 0.$

This equation is a cubic in k^2, and is satisfied by $k^2 = 4$, or $k = \pm 2$.

Putting $k = 2$, we have

$$\left.\begin{array}{l} m + l = 2 \\ m - l = 4 \end{array}\right\}$$

i.e., $m = 3, l = -1$.

Thus, $x^4 - 2x^2 + 8x - 3 = (x^2 + 2x - 1)(x^2 - 2x + 3)$,

so that $x^2 + 2x - 1 = 0$, and $x^2 - 2x + 3 = 0$ give the roots of the biquadratic as $-1 \pm \sqrt{2}, 1 \pm \sqrt{-2}$

78. How about solving the general equation of the fifth degree ?

After the solutions of the cubic and the biquadratic equations were obtained, many gifted mathematicians continued their efforts to crack the fifth degree equation. Their efforts spread over a period of more than two centuries and a half but without any success in sight.

79. Is it not possible to solve a fifth degree equation by making its solution depend upon a suitable fourth degree equation ?

We have seen that the solution of an equation depends on the solution of a lower degree equation. Using this principle, a French mathematician, Lagrange, tried to solve the fifth degree equation but it led him to an equation of the sixth degree. This was as indirect indication that a general equation of the 5th degree cannot be solved by such methods. Lagrange missed the hint.

80. What did Abel prove ?

Abel, a Norwegian mathematician, in 1824 proved the remarkable result that the general algebraic equation of degree higher than the fourth is unsolvable by radicals, i.e., root extractions.

81. But some equations of degree greater than four like $x^6 - 1 = 0$ can be completely solved by root extraction !

Equations like $x^6 - 1 = 0$, $x^8 - 2 = 0$, $x^n - a = 0$ can be completely solved by root extraction though each is of degree greater than four. Not only these, but there are also many special equations of arbitrary degree, which can be solved by root extraction, so the problem now was to determine the precise conditions for the solvability of an equation in radicals.

82. Who determined these precise conditions ?

A french mathematician Galois, who died in a senseless duel at the young age of 21, went deep into the problem and showed in 1831 that an algebraic equation is solvable by radicals if and only if its Galois group is solvable. The proof is too difficult to be given here.

83. When are approximation methods used ?

Although a general equation of degree greater than four cannot be solved by root extraction, the roots of any equation with numerical coefficients can, however, be found to any degree of accuracy by what are called approximation methods.

Many methods are available and different methods are suitable for different equations.

84. Are these methods suitable for the cubic and the biquadratic ?

Such methods are more suitable for solving the cubic and the biquadratic equations with numerical coefficients.

85. How is a cubic solved by such method ?

The method adopted here is suitable if the given equation can be reduced to the form :

$$x = a + \phi(x),$$

where a is some number and $\phi(x)$ is a small quantity depending upon x.

An approximate root is given by $x = a$.

Putting $x = a$ in the right hand side of the given equation, we get a second approximation,

$x = a + \phi(a)$, where $\phi(a)$ means a substituted for x in $\phi(x)$.

Denoting this value by a_1, a third approximation is

$$x = a + \phi(a_1)$$

and so on till the root to the required degree of accuracy is obtained.

86. How is the following equation solved : $x^3 + 3x^2 + 2 = 0$?

On division by x^2 the given equation may be written as

$$x = -3 - \frac{2}{x^2}, \text{ which is of the form } x = a + \phi(x)$$

A first approximation is $x = -3$.

Second is $x = -3 - \frac{2}{(-3)^2}$

$= -3 - \frac{2}{9}$

$= -3.22$

Third is $x = -3 - \frac{2}{(-3.22)^2}$

$= -3.193$

Next is $x = -3 - \frac{2}{(-3.193)^2}$

$= -3.1962$

This equation has only one real root. The other two roots are imaginary.

87. How are the roots of an equation located ?

When an equation has more than one real roots, to determine all the roots it is necessary to locate them approximately before their value can be determined to the required degree of accuracy.

Consider the equation.

$$8x^3 - 100x^2 + 342x - 315 = 0$$

Give to x the values 0, 1 2, 3, 4, 5, etc. and commute the values of the expression on the left. Let it be called P, then

$$P = 8x^3 - 100x^2 + 342x - 315$$

When $x = 0$ $P = -315$

$x = 1,$ $P = -65$

$x = 2,$ $P = 33,$

$x = 3,$ $P = 27,$

$x = 4,$ $P = -35,$

$x = 5,$ $P = -105,$

$x = 6,$ $P = -135,$

$x = 7,$ $P = -77,$

$x = 8,$ $P = 120,$

From the above we see that when x increases from 1 to 2, P increases from – 65 to 33. Starting from a negative value – 65, P must first attain a zero value, and then only it can increase to the positive value 33. P will, therefore, attain a zero value for some value of x between 1 and 2.

Similarly, when x increases from 3 to 4, value of P decreases from 27 to – 35. P will, therefore, again attain a zero value for some value of x between 3 and 4.

Next, when x increases from 4 to 7, P retains the same sign and does not attain a zero value in between.

Finally, when x increases from 7 to 8, value of P increases from – 77 to 120. P will, therefore, again attain a zero value for some value of x between 7 and 8.

Since a zero value of P corresponds to a root of the equation, the given equation has roots between 1 and 2, 3 and 4, and 7 and 8 only.

In the present case, P attains zero values for x = 1.5, 3.5, and 7.5, which are the roots of the given equation.

88. What are simultaneous equations ?

When two or more equations are satisfied by the same values of the unknown quantities, they are called simultaneous equations.

An example of simultaneous equations containing two unknowns, x and y, is

$$3x + 4y = 18,$$

$$5x + 7y = 31.$$

89. How are they solved ?

One solves such equations at school by first eliminating either x or y.

Here y can be eliminated by multiplying the first equation by 7, and the second by 4, and subtracting. Thus

7 times the I equation gives : $21x + 28y = 126,$

4 times the II equation gives : $20x + 28y = 124,$

Subtracting, $x = 2.$

Substituting the value of x in any one of the two given equations, say in the first, we get

$6 + 4y = 18,$

or $4y = 12,$ or $y = 3.$

Thus the solution is $x = 2, y = 3.$

90. How are simultaneous equations containing three unknowns solved ?

The method for solving simultaneous equations containing more than two unknowns is similar.

Example :

$$x + y + 3z = 12,$$
$$2x + 3y + 4z = 20,$$
$$3x + 2y + 5z = 22,$$

First z is eliminated from the first two equations, and then from the last two equations. This gives us two equations in only two unknowns, x and y. These can be solved as usual.

Thus,

4 times I equation gives : $4x + 4y + 12z = 48,$

3 times II equation gives : $6x + 9y + 12z = 60.$

Subtracting, $-2x + 5y = -12,$

or $2x + 5y = 12$ .....(A)

Again,

5 times II equation gives : $10x + 15y + 20z = 100,$

4 times III equation gives : $12x + 8y + 20z = 88,$

Subtracting, $-2x - 7y = 12,$

or $2x - 7y = -12$ .....(B)

Solving (A) and (B) as usual, we get $x = 1, y = 2.$

Substituting the values of x and y in the first equation, we get $z = 3.$

Thus we have $x = 1,\ y = 2,\ z = 3.$

91. What is meant by indeterminate equations ?

If the number of unknowns is greater than the number of equations, the equations are said to be indeterminate.

Such equations have an unlimited number of solutions.

For example, consider the equation $3x + y = 10.$

It can be written as $y = 10 - 3x$

Here, corresponding to any value of x, y has a value.

Thus the equation has an unlimited number of solutions.

But if the equation is to be solved only in positive integers, the number of solutions becomes limited.

92. How to solve in positive integers the equation : $3x + y = 10$?

The given equation can be written as

$$y = 10 - 3x$$

Since y must be a positive integer, x can assume only the values 0, 1, 2, or 3. If x be given a value greater than 3, y becomes negative.

The following are, therefore, the positive integral solutions of the given equation.

$$x = 0, y = 10;$$
$$x = 1, y = 7;$$
$$x = 2, y = 4;$$
$$x = 3, y = 1$$

93. What are Diophantine equations ?

Indeterminate equations are also known as Diophantine equations in honour of the ancient Greek mathematician Diophantus (250 A. D.), who was the first to systematically discuss such equations, and showed extraordinary skill in solving them.

94. What gives rise to such equations and how are they solved ?

Problems of the following type lead to indeterminate equations.

Problem : A person spends Rs. 414 in buying pens and pencils. If each pen costs Rs. 13 and each pencil Rs. 11, how many of each does he buy ?

Let x be the number of pens and y that of pencils, then

$$13x + 11y = 414, \quad .....(1)$$

where x and y are positive integers.

The following is the method of solution :

Divide throughout by 11, the smaller of the two coefficients, then,

$$x + \frac{2}{11}x + y = 37 + \frac{7}{11},$$

or $$x + y + \frac{2x - 7}{11} = 37.$$

Since x and y are integers, we must have

$$\frac{2x-7}{11} = \text{Integer.}$$

Now we multiply 2x – 7 by an integer such that the coefficient of x differs by unity from 11 or a multiple of 11.

Such an integer in this case is 6.

[It is necessary to employ a similar artifice before introducing a symbol for the integer].

Multiplying 2x – 7 by 6, we have

$$\frac{12x-42}{11} = \text{Integer,}$$

or $$x + \frac{x}{11} - 3 - \frac{9}{11} = \text{Integer,}$$

or $$x - 3 + \frac{x-9}{11} = \text{Integer,}$$

i.e., $$\frac{x-9}{11} = \text{Integer} = p \text{ (suppose),}$$

therefore $$x = 11p + 9. \qquad \text{....(2)}$$

Substituting this value of x in (1),

$$y = 27 - 13p. \qquad \text{....(3)}$$

From (3) we see that if p is greater than 2, y becomes negative. The integral positive values of x and y are therefore obtained by putting p = 0, 1 and 2 only.

Thus the complete solution is given by :

$$\left.\begin{array}{l} p=0,\ x=9,\ \ y=27; \\ p=1,\ x=20,\ y=14; \\ p=2,\ x=31,\ y=1. \end{array}\right\}$$

95. What is Bnaskar's broken bamboo problem ?

Bhaskaracharya, the famous Hindu mathematician has given this problem in his well–known treatise, the Lilavati.

It appears in the following form : If a bamboo 32 cubits high is broken by the wind so that the tip meets the ground 16 cubits from the base, at what height above the ground was it broken ?

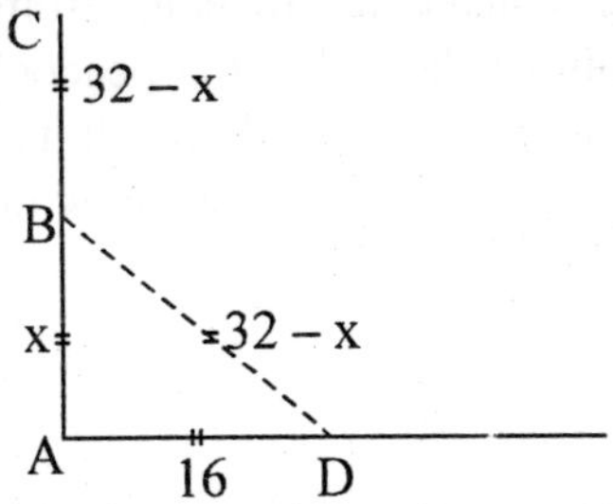

The Theorem of Pythagoras is made use of in the solution.

Let the bamboo AC be broken at a height x cubits above the ground so that

AB = x, BC = 32 – x = BD, AD = 16.

Then by the Theorem of Pythagoras,

$$AB^2 + AD^2 = BD^2,$$

or $x^2 + 16^2 = (32 - x)^2,$

or $x^2 + 256 = 1024 - 64x + x^2,$

or $64x = 768$ or $x = 12$ cubits.

The bamboo was broken at a height 12 cubits above the ground. (A cubit is an old measure of length 18 to 22 inches.)

To verify :–

$$AB + BC = 12 + 20 = 32.$$

$$AB^2 + AD^2 = BD^2,$$

$\therefore \quad 12^2 + 16^2 = 20^2.$

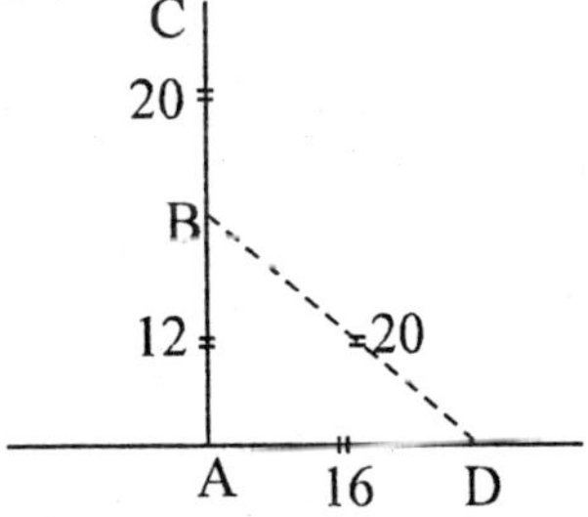

96. What is Bhaskar's peacock- and- the snake problem ?

It is also given by Bhaskaracharya in his book Lilavati. It also makes use of the Theorem of Pythagoras, but it leads to an indeterminate equation.

It appears in the following form :

A peacock is perched atop a pillar at the base of which is a snake's hole. Seeing a snake at a distance from the pillar which is three times the height of the pillar, the peacock pounces upon the snake in a straight line before it could reach its hole. If the peacock and the snake had gone equal distances, how many cubits from the hole did they meet ?

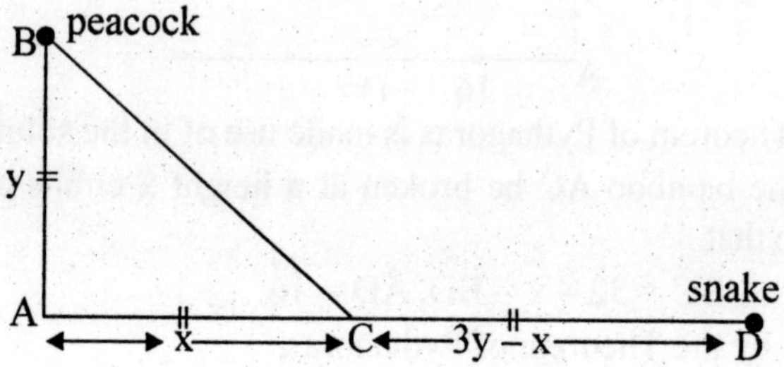

Let the snake's hole be at A. Let AB be the pillar, let the peacock be at B, and the snake at D.

Let them meet at C a distance x cubits from the hole. Let y be the height of the pillar, so that.

$$AC = x, AB = y, AD = 3y \text{ (given)}$$

and $CD = 3y - x = BC$, $\because BC = CD$ (given)

Then by the Theorem of Pythagoras,

$$AB^2 + AC^2 = BC^2,$$

or $y^2 + x^2 = (3y - x)^2$,

or $y^2 + x^2 = 9y^2 - 6xy + x^2$,

or $8y^2 - 6xy = 0$,

or $2y(4y - 3x) = 0$.

Dividing by 2y, we get,

$$4y - 3x = 0,$$

or $x = \frac{4}{3}y$.

This is an indeterminate equation having many solutions.

Some of them are :

If $y = 3, x = 4$;

If $y = 6, x = 8$;

If $y = 9, x = 12$; etc.

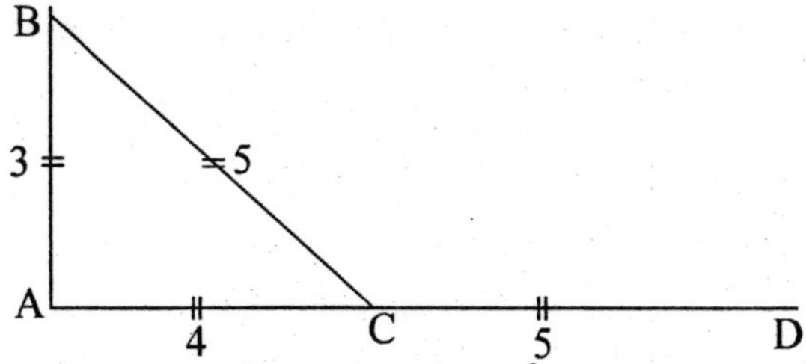

97. De Morgan, who lived in the 19th century, proposed the following conundrum concerning his age :

I was x years old in the year x^2. When was I born ?

A conundrum is a puzzling question like a riddle.

Square years can be spotted as follows. Beginning with 40, we have

$$40^2 = 1600$$
$$41^2 = 1681$$
$$42^2 = 1764$$
$$\boxed{43^2 = 1849}$$
$$44^2 = 1936$$

Since De Morgan lived in the 19th century, i.e., in 1801-1900, he was 43 years old in the year 1849, and was, therefore, born in the year 1806.

98. How are simultaneous indeterminate equations solved ?

When two indeterminate equations in 3 unknowns are given, any one unknown, say z, is eliminated, and a single indeterminate equation in two unknowns is obtained.

The resulting equation is solved as usual.

Consider the following problem:

The expenses of a party numbering 44 were Rs. 451. If each man paid Rs. 15, each woman Rs. 12, and each child Rs. 5, how many were there of each ?

Let x, y, z denote the number of men, women and children respectively, then we have

$$x + y + z = 44, \qquad .....(1)$$
$$15\,x + 12\,y + 5\,x = 451. \qquad .....(2)$$

Multiplying first equation by 5 and subtracting from the second to eliminate z, we get,

$$10x + 7y = 231. \qquad .....(3)$$

This is an equation in two unknowns and can be solved for positive integral values in the usual way.

Dividing by the smaller coefficient 7, we get

$$x + \frac{3x}{7} + y = 33.$$

Since x and y are integers,

$$\frac{3x}{7} = \text{Integer},$$

or $$\frac{15x}{7} = \text{Integer},$$

or $$2x + \frac{x}{7} = \text{Integer},$$

or $$\frac{x}{7} = \text{Integer} = p \text{ (suppose)},$$

$$\therefore \quad x = 7p.$$

Substituting this value of x in (3), we get $y = 33 - 10p$.

Substituting for x and y in (1), we get $z = 3p + 11$.

Now, p can be given the values 1, 2 and 3 only, because greater values of p make y negative.

The complete solution is thus given by :

$p = 1, x = 7, y = 23, z = 14$;

$p = 2, x = 14, y = 13, z = 17$;

$p = 3, x = 21, y = 3, z = 20$.

99. How is the following equation solved in positive integers : $2xy - 4x^2 + 12x - 5y = 11$?

The given equation can be written as :

$$2xy - 5y = 4x^2 - 12x + 11,$$

or $$(2x - 5)y = 4x^2 - 12x + 11.$$

Expressing y in terms of x, we get

$$y = \frac{4x^2 - 12x + 11}{2x - 5},$$

or $$y = 2x - 1 + \frac{6}{2x-5}, \text{ on division by } 2x - 5.$$

For y to be an integer, $\frac{6}{2x-5}$ must be in integer.

Since ± 1, ± 2, ± 3 and ± 6 are the only divisors of 6,

$\therefore$ $2x - 5 = \pm 1, \pm 2, \pm 3$ and ± 6.

Out of these $2x - 5 = \pm 2$ and ± 6 do not yield integral values of x, and must be rejected.

$2x - 5 = \pm 1$ and $2x - 5 = \pm 3$ give $x = 3, 2, 4$ and 1.

These values give :

$x = 3, y = 11$; $x = 2, y = -3$;

$x = 4, y = 9$; $x = 1, y = -1$.

Out of these the admissible solutions are

$x = 3, y = 11$; $x = 4, y = 9$.

100. What is the general indeterminate equation of the second degree ?

The second degree equation : $Ny^2 + 1 = x^2$,

where N is a positive integer but not a perfect square, is known as the general indeterminate equation of the second degree.

It can always be solved in positive integers, the number of solutions being unlimited.

The method of solution is difficult and not suitable to be given here.

101. What is remarkable about the equation $61y^2 + 1 = x^2$?

This is a particular case of the general equation referred to above, where N is given the value 61.

Bhaskaracharya, the great Hindu mathematician, is famous for obtaining general integral solutions of this equation by what is known as the "cyclic method".

To illustrate the method, he, in his book 'Bija ganita' wirtten in 1150 A.D. gave a worked out example : $61y^2 + 1 = x^2$.

What is remarkable is that this very problem was proposed for solution about 500 years later by the eminent French mathematician Fermat to his friend Frenicle in 1657.

But it was solved by Euler as late as 1732.

Bhaskaracharya gave the following solution :

$x = 1,776,319,049,$

$y = 22,615,390.$

102. How is the non-algebraic equation $xe^x = 2$ solved ?

It is best solved graphically :

The single given equation can be expressed in the form of two separate equations :

$$y = e^x, \text{ and } y = \frac{2}{x}$$

These two together are equivalent to the single given equation $x\,e^x = 2$.

The x coordinate of the point of intersection of the two curves will give the required value of x.

To plot the curve $y = e^x$, we use the values :

when $x = 0, \quad y = 1$

$x = 1, \quad y = e^1 = 2.7,$

$x = 2, \quad y = e^2 = 7.3$

To plot the curve $y = \frac{2}{x}$, we use the values :

when $x = .5, \quad y = 4,$

$x = 1, \quad y = 2,$

$x = 2, \quad y = 1,$

$x = 3, \quad y = .6,$

$x = 4, \quad y = .5,$

The two curves are plotted in the same figure below :

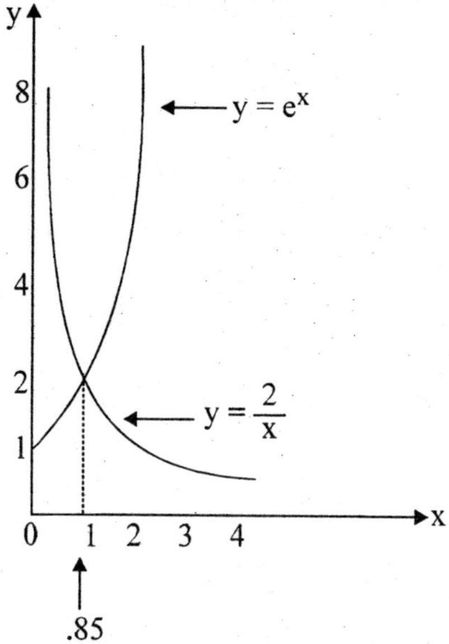

From the figure, x coordinate of the point of intersection is.85

$\therefore$ x =.85 approximately.

103. Abstract Algebra again ! How is modern abstract algebra different from classical algebra ?

Classical algebra is a generalisation of arithmetic, while abstract algebra is a study of algebraic structures.

Algebraic structures are also known as algebraic systems.

104. What precisely is meant by an algebraic system ?

An algebraic system is a SET of objects called elements, together with one or more operations for combining them.

105. What is meant by a SET ?

By a SET is meant any well–defined collection of objects.

The elements of a set are separated by commas and written within curly brackets.

For example, the set of first five natural numbers is written as $A = \{1, 2, 3, 4, 5\}$, where A is a name given to the set.

Sets are often denoted by single capital letters A, B, C, X. etc.

The elements are generally denoted by lower case letters a, b, c,........ x, y, z.

106. What is meant by an operation ?

A rule of combination of two elements is called an operation.

Addition and multiplication are the most familiar examples of an operation.

107. What can be the elements and operations of a set ?

The elements are not necessarily numbers and the operations not necessarily those of arithmetic.

When the operations are not necessarily addition and multiplication, they are denoted by the symbols 0 and *, spoken as 0 and star respectively. o is the first letter of the word operation.

108. What are the most commonly used algebraic systems ?

The most commonly used algebraic systems are GROUPS, RINGS, INTEGRAL DOMAINS, FIELDS and VECTOR SPACES.

The classification depends upon the axioms that are satisfied under the particular operation or operations defined for the elements of the set.

109. What is meant by the closure property ?

The foremost property common to all operations is that the operation applied to any two elements of the set produces an element which also belongs to the set. This is known as the closure property and the system is said to be closed under that operation.

For example, consider the set of integers :

$I = \{..........., -3, -2, -1, 0, 1, 2, 3, ...\}$,

where the dots mean that the sequence of the positive as well as the negative integers is to be extended indefinitely on both the sides.

The set of integers is said to be closed under addition and multiplication, because the sum of any two integers is an integer, and the product of any two integers is also an integer.

Symbolically, if a and b are the elements of a set, then to satisfy the closure property, the element a o b should also be an element of the same set.

The symbol o for operation is noncommittal in that it can be addition, multiplication or any other operation whatever.

110. What is meant by the associative property ?

If a, b, c are the elements of a set, then the associative property is said to hold good if

a o (b o c) = (a o b) o c.

The brackets here simply mean that the elements within them are to be dealt with first.

For example, if the operation o stands for addition, and the elements be numbers, then the numbers within the brackets are to be added first.

Thus, $2 + (3 + 4) = (2 + 3) + 4$ means that

$2 + 7 = 5 + 4$ must be true.

111. What is a GROUP ?

A set G of elements a, b, c,....., on which an operation o is defined, is said to form a GROUP with respect to the operation if the following properties hold for every a, b, c,......in G :

(i) a o b is in the set G (Closure property),

(ii) a o (b o c) = (a o b) o c (Associative property),

(iii) There exists a special element, called the identity element e in the set G such that e o a = a for every a in G (Existence of identity element),

(iv) For every element a in G there exists an element a^{-1} in G such that a o a^{-1} = e (Existence of inverses).

112. Does the set of integers with respect to addition form a group ?

Here the operation o stands for ordinary addition, i. e., +.

(i) Since the sum of any two integers is an integer, the closure property is satisfied.

(ii) That any three integers a, b, c added together in the two sequences result in the same sum shows that the associative law is satisfied.

(iii) The identity element for addition is o, i. e., zero, so that a + o = a is a statement true for every integer a.

(iv) Each element a has an additive inverse – a, which is also an integer. For example, the additive inverse of 2 is – 2, and 2 + (– 2) = 0, where o is the identity element.

Thus the set of integers satisfies all the four properties so that it forms a group with respect to addition.

113. Do the integers form a group with respect to multiplication ?

(i) The product of any two integers is an integer, therefore the closure property is satisfied.

(ii) That any three integers a, b, c multiplied in the two sequences result in the same product shows that the associative law is satisfied.

(iii) The identity element for multiplication is 1, which is an integer, so that the third property is also satisfied.

(iv) The fourth property is not satisfied, because the multiplicative inverse of an integer is not an integer, but a fraction (except when the integer is 1 or – 1).

As an example, $2 \times \frac{1}{2} = 1$, so that the multiplicative inverse of the integer 2 is the fraction $\frac{1}{2}$ and not an integer.

Since one of the four requirements of a group is not satisfied, the integers do not form a group with respect to multiplication.

114. What is an Abelian group ?

A group G is said to be an Ablelian or commutative group, if one more property of commutativity of the operation is also satisfied, i. e., a o b = b o a for every a, b in G.

The integers with respect to addition form an Abelian group, since a + b = b + a is true for all integers.

Commutativity implies that the order of elements in the operation does not make any difference.

Since 2 + 3 = 3 + 2, and this property is true for any two integers, addition is said to be commutative for the set of integers.

115. What entities can be the "elements" of a group and what kind of operations can be defined on them ?

The elements of a group can be the numbers as in arithmetic, or points as in geometry, They can be transformations as in algebra or geometry, or anything at all.

The operation can be addition or multiplication as in arithmetic. It can be a rotation about a point or an axis as in geometry. It can be any other rule for combining two elements of a set to form a third element in the set, as in the case of two transformations applied one after the other to give a third transformation.

116. What led to the theory of groups ?

Originally, the theory of groups was developed to examine why some equations of degree higher than four could be solved by root extraction while others could not be so solved.

In such enquiry only, it was first observed that the symmetry of the roots of the equation was fundamental for the solution of the whole problem.

Later the theory was utilized as an instrument for investigating important regularities, like symmetries, of the real world.

117. What purpose is served by studying symmetries ?

The examination of various symmetries leads to deep insights which cannot otherwise be gained by direct inspection.

Group theory helps to reveal the symmetries, which in turn help in the deeper understanding of various phenomena.

118. Where else is the theory of groups applied ?

It is applied in many branches of science, notably in geometry, crystallography, physics, chemistry, molecular biology, and space and time theory, where the laws of symmetry find an important role.

119. What is isomorphism ?

It may so happen that two groups with quite different types of elements and operations may be essentially the same in an abstract sense.

For example, consider the group $\{I, R, R^1\}$ of the rotations of an equilateral triangle through 0°, 120°, 240° respectively, and the group $\{1, \omega, \omega^2\}$ of the cube roots of unity with respect to multiplication.

Following are the multiplication tables for the two groups :

	I	R	R′
I	I	R	R′
R	R	R′	I
R′	R′	I	R

	1	ω	ω^2
1	1	ω	ω^2
ω	ω	ω^2	1
ω^2	ω^2	1	ω

First table expresses that :

The elements of the top row, when multiplied by I give I, R, R′; when multiplied by R give R, R′, I; when multiplied by R′ give R′, I, R.

Similarly for the other table.

It may be noticed that the first table becomes identical with the second if I, R, R′ are respectively replaced by 1, ω, ω^2.

The two groups show a one to one correspondence between the respective elements, and their products, and are called isomorphic to each other.

Two isomorphic groups are said to be abstractly identical.

120. What purpose is served if two different groups are known to have the same structure ?

It two different groups are known to have the same structure, the familiar one is worked out and the results are then interpreted in terms of the less familiar.

The theory of groups thus reveals the identify behind apparent dissimilarities, and opens the way to a treasure of information otherwise impossible to obtain.

121. How is it that group theory has split into several branches ?

The methods and concepts of the theory of groups proved to be extremely important not only for the investigation of the laws of symmetry but also for the solution of many other problems.

Since every domain of application has its own particular problems, the growing number of these domains necessitated the creation of new branches of the theory of groups suitable for different themes and contexts.

Thus the theory of groups, which is a single entity in its essential concepts, split into a number of more or less independent disciplines.

122. Which are these split branches ?

Some of them are the following :

The general theory of groups,

The theory of finite groups,

The theory of continuous groups,

The theory of diserete groups of transformations,

The theory of representations and characters of groups.

123. What is a RING ?

A ring is a system with two operations.

A set R of elements on which two operations + and. (called addition and multiplication) are defined, is said to be a ring if

(i) Addition satisfies closure, associative and commutative properties,

(ii) R contains additive inverse for each element and also contains the identity element zero.

(iii) Multiplication satisfies closure and associative properties, and is distributive over addition, i.e., a.(b + c) = a. b + a. c, and (b + c). a = b. a + c. a for all a, b, c in R.

Briefly, a ring

(1) forms a commutative group with respect to addition, and

(2) The operation called multiplication satisfies the closure, associative and distributive properties.

124. What is a familiar example of the ring ?

Under addition and multiplication of integers, the set of integers is a ring.

125. What is a commutative ring ?

If the operation called multiplication is also commutative, i.e., a.b = b. a is true for all the elements, then the ring is called a commutative ring.

126. But is multiplication not always commutative ?

No, not always.

It is common knowledge that 4 × 5 = 5 × 4, i. e., a × b = b x a.

But this is true for such familiar systems only as integers, fractions, rational numbers, etc.

There are systems for which this is not true.

Vectors and matrices are the examples.

127. What is a Vector ?

Any line segment with which is also associated a direction is called a vector.

Thus a straight line from O to P is a vector, whose length is O P and whose direction is from O to P.

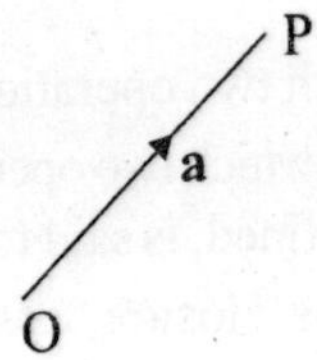

It is denoted by $\overrightarrow{OP}$ or any small case bold letter, say **a**.

If $\overrightarrow{OP} = \mathbf{a}$ and $\overrightarrow{PR} = \mathbf{b}$, then $\overrightarrow{OR}$ is called the vector sum or resultant of the vectors **a** and **b**.

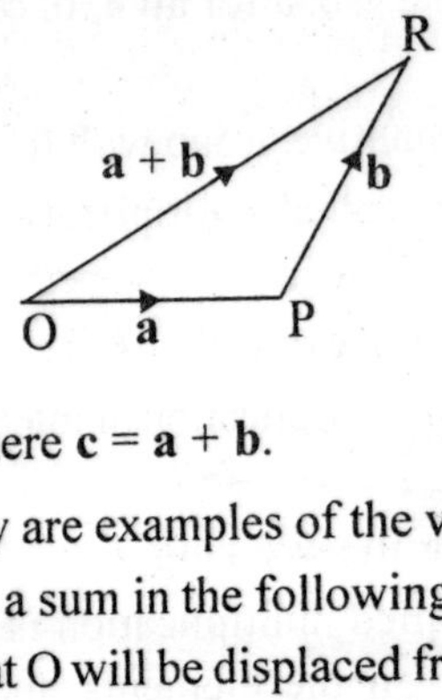

Thus $\overrightarrow{OR} = \mathbf{c}$, where $\mathbf{c} = \mathbf{a} + \mathbf{b}$.

Force and velocity are examples of the vector quantities.

The vector sum is a sum in the following sense :

A particle placed at O will be displaced from the position O to the position P by the vector **a**, and the vector **b** will further displace the particle from P to R. Thus the ultimate displacement is from O to R. It is in this sense that $\overrightarrow{OR} = \overrightarrow{OP} + \overrightarrow{PR} = \mathbf{a} + \mathbf{b}$.

Again, we may add to this any other vector

$\mathbf{c} = \overrightarrow{RS}$, and obtain the result $\overrightarrow{OS} = \overrightarrow{OR} + \overrightarrow{RS}$.

$$= (\mathbf{a} + \mathbf{b}) + \mathbf{c}$$
$$= \mathbf{a} + \mathbf{b} + \mathbf{c}.$$

This type of addition holds for any number of vectors.

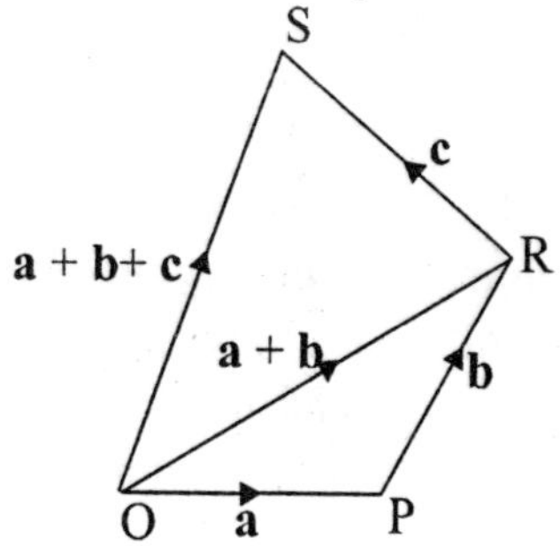

128. In what way is the product of two vectors not commutative ?

For vectors, two kinds of product are defined.

One is called scalar product or dot product. It is commutative so that for any two vectors **a** and **b**, we have **a. b** = **b. a**.

The other is called vector product or cross product. It is not commutative, because for any two vectors **a** and **b**, we have.

$(\mathbf{a} \times \mathbf{b}) = -(\mathbf{b} \times \mathbf{a}).$

129. What is a matrix ?

An arrangement of numbers in a rectangular form is called a matrix.

Thus $A = \begin{bmatrix} 2 & 1 & -1 \\ 1 & 3 & 2 \end{bmatrix}$

is a matrix having 2 rows and 3 columns, and is called a matrix of order 2 × 3.

A matrix having m rows and n columns is called a m × n matrix (read as m by n matrix).

130. In what way is the product of two matrices not commutative?

In the case of matrices also, the product of two matrices is, usually, not commutative.

For example, it A be the matrix $\begin{bmatrix} 1 & 2 \\ 3 & 4 \end{bmatrix}$, and B the matrix $\begin{bmatrix} 5 & 6 \\ 7 & 8 \end{bmatrix}$, the products AB and BA will be different.

Thus

$$AB = \begin{bmatrix} 1 & 2 \\ 3 & 4 \end{bmatrix}\begin{bmatrix} 5 & 6 \\ 7 & 8 \end{bmatrix} = \begin{bmatrix} 1\times5+2\times7 & 1\times6+2\times8 \\ 3\times5+4\times7 & 3\times6+4\times8 \end{bmatrix} = \begin{bmatrix} 19 & 22 \\ 43 & 50 \end{bmatrix},$$

While

$$BA = \begin{bmatrix} 5 & 6 \\ 7 & 8 \end{bmatrix}\begin{bmatrix} 1 & 2 \\ 3 & 4 \end{bmatrix} = \begin{bmatrix} 5\times1+6\times3 & 5\times2+6\times4 \\ 7\times1+8\times3 & 7\times2+8\times4 \end{bmatrix} = \begin{bmatrix} 23 & 34 \\ 31 & 46 \end{bmatrix}.$$

131. Where is the matrix theory applied ?

It has been found of great utility in many branches of higher mathematics, such as algebraic and differential equations, astronomy, mechanics, theory of electrical circuits, quantum mechanics, nuclear physics and aerodynamics.

132. What is an INTEGRAL DOMAIN ?

A commutative ring is said to be an integral domain, if

(i) it has unity element,

(ii) it is without zero divisors.

133. What is meant by zero divisors ?

By zero divisors is meant non–zero elements a, b such that their product ab = 0.

An example of such divisors is met with in what is called modular arithmetic or arithmetic modulo n.

For a give integer n, such arithmetic is obtained by using only the integers 0. 1, 2,....., n – 1.

Addition and multiplication are defined as the remainder after division by n of the ordinary sum and product of any two integers of the system.

For example, if n = 6, then

$2 + 4 \equiv 0$, $2 + 5 \equiv 1$, $3 + 5 \equiv 2$, because

2 + 4 i.e. 6, divided by 6 leaves the remainder 0,

2 + 5 i.e. 7, divided by 6 leaves the remainder 1,

and 3 + 5 i.e. 8, divided by 6 leaves the remainder 2,

Similarly, $2 \times 3 \equiv 0$, $4 \times 5 \equiv 2$.

2 and 3 are called zero divisors of the system called "integers modulo 6", because their product is congruent to zero, but neither 2 nor 3 is equal to zero.

The symbol ≡ is read as 'congruent to'.

If n is prime, there are no zero divisors.

134. In a certain mathematical system, 2 × 2 = 2 × 5. What is the system called ?

The system is called "integers modulo 6," in which 4 is congruent to 10, modulo 6,

or $4 \equiv 10 \pmod 6$.

135. What is a FIELD ?

A ring is called a field if

(i) it is commutative,

(ii) it has unity element,

(iii) its every non–zero element has multiplicative inverse,

(iv) it has no zero divisors.

136. In what way are the integral domain and the field distinct from a ring ?

The important difference is that the ring can have zero divisors whereas the integral domain and the field cannot have such divisors.

137. What is a VECTOR SPACE ?

Let V = {x, y, z,......} be a set, and let α be an element in a field F, and let the product α x be defined for every element α in the field F, and every element x in the set V.

Then V is said to form a vector space over F if the following properties hold :

1. V forms an abelian group with respect to addition,
2. For every x, y in V, and α, β in F,
   (i) α x is in V (closure property),
   (ii) $\alpha (\beta x) = (\alpha \beta) x$ (Associative law),
   (iii) $\alpha (x + y) = \alpha x + \alpha y$ (Distributive law),
   (iv) $(\alpha + \beta)x = \alpha x + \beta x$ (Distributive law),
   (v) $1x = x$ (multiplication by unity, where 1 is the unity element of F).

The elements of V are called vectors, and those of F scalars.

Note—In dealing with a vector space V over F, we have four compositions, viz.,

(i) Addition in V,

(ii) Scalar multiplication of the elements of V with the elements of F,

(iii) Addition in F,

(iv) Multiplication in F.

138. In what way is a vector space different from an integral domain ?

In an integral domain the operation of multiplication is defined for the elements of its set, but in the vector space multiplication for the elements of V is not defined.

In vector space, however, scalar multiplication of the elements of the set V by the elements of the field is defined.

139. How does a vector space differ from other algebraic structures?

Groups are algebraic systems with one operation.

Rings, integral domains and fields are algebraic systems with two operations.

A vector space combines two different algebraic systems into a single system.

140. Where is the vector space structure used ?

A large number of models arising in the solution of specific problems in the field of Physics, Chemistry, Economics and other social sciences is found to possess a vector space structure.

The key to their study, therefore, lies with the study of vector spaces.

141. What are Hilbert space and Banach space ?

We have seen that a vector space is a set of elements called vectors, subject to the usual rules of combination of vectors and scalars.

Vector spaces of varying types are determined according to the restrictions imposed on the vector elements.

Hilbert space, for example, is a vector space whose elements have infinitely many components, subject to the condition that $(x_1^2 + x_2^2 + \ldots\ldots)$ be finite.

Hilbert space on account of its extraordinary degree of abstraction has found application in quantum theory.

Banach space is a still more abstract vector space in which the elements need not be defined with respect to the real or the complex number field.

The Banach space is a real Banach space or a complex Banach space according as the scalar multipliers are real numbers or complex numbers.

Hilbert space is an example of the Banach space.

142. What is Linear Algebra ?

A vector space A (over field F) is called an algebra or a linear algebra if its vectors can be multiplied in such a way that

(i) $x(yz) = (xy)z$, (Associative property)

(ii) $\left.\begin{array}{l} x(y+z)=xy+xz, \text{and} \\ (x+y)z=xz+yz, \end{array}\right\}$ (Distributive property)

(iii) $\alpha(xy) = (\alpha x)y = x(\alpha y)$,

where α is any scalar in F, and x, y, z vectors in A.

By scalar is meant real as well as complex numbers.

143. Which operations are used in Boolean Algebra ?

In Boolean Algebra, the familiar operations of addition and multiplication are replaced by union and intersection, denoted by the symbols $\cup$ and $\cap$ respectively.

144. What are they ?

If A and B be two sets of integers, say,

$A = \{1, 2, 3, 4, 5\}$, and

$B = \{4, 5, 6, 7, 8\}$,

then $A \cup B$ is defined as the set containing all the elements which are in A or B (or in both), so that

$A \cup B = \{1, 2, 3, 4, 5, 6, 7, 8\}$.

$A \cap B$ is defined as the set of all elements common to both A and B, so that, here,

$A \cap B = \{4, 5\}$.

145. What are these : Subset, Null set, Universal set, and Complement of a set ?

Subset – A set A is a subset of the set B (or is contained in B) if each element of A is also an element of B.

Thus, if A = {1, 2, 3} and B = {1, 2, 3, 4, 5}, then A is called a subset of B.

Null set – If, however, no element is common to A and B, $A \cap B$ is called an empty set or null set and is denoted by the symbol ϕ.

Thus, if A = {1, 2, 3} and B = {4, 5, 6}, then $A \cap B = \phi$.

Universal set – If, in any discussion, many sets are involved in the context of a single fixed set, this set is called the *Universal set*, and is denoted by U.

The other sets are called subsets of U.

Complement of a set – The complement of a set, denoted by A′, is the set of all elements which are not in A.

Since the only elements we consider are those which make up U, A′ consists of all those elements in U which are not in A.

Thus we have $A \cup A' = U$, and $A \cap A' = \phi$.

146. What is Boolean Algebra ?

A Boolean Algebra consists of a set of elements together with two operations, denoted by symbols $\cup$ and $\cap$.

These operations are defined on the set and satisfy the following axioms :

(i) $\cup$ and $\cap$ are both commutative operations, i.e., $a \cup b = b \cup a$, and $a \cap b = b \cap a$,
a, b being any elements of the set.

(ii) Each of the operations $\cup$ and $\cap$ distributes over the other, i. e.,
$a \cup (b \cap c) = (a \cup b) \cap (a \cup c)$,
and$a \cap (b \cup c) = (a \cap b) \cup (a \cap c)$,
a, b, c being any elements of the set.

(iii) Identily elements z and u exist in the set such that $a \cup z = a$, and $a \cap u = a$, where $z \neq u$, for each element a of the set.

(iv) For each element a in the set, there exists a distinct element a′, called the complement of a, with the property that $a \cup a' = U$, and $a \cap a' = 0$.

(v) And, of course, the closure condition must be satisfied, which means that the operations of union, intersection and complementation on the elements of the set result in the elements of the set only.

147. What can be the elements of the set of a Boolean Algebra ?

The elements of the set may be abstract objects, or concrete things such as numbers, propositions, sets, or electrical networks.

Boolean algebra can be interpreted as an algebra of propositions as well as an algebra of sets.

148. Is there only one Boolean Algebra ?

No, one could expect several variations depending upon what the elements of the set are, and what the operations.

The operations denoted by the symbols $\cup$ and $\cap$ are not necessarily the union and the intersection. They are basically undefined operations satisfying only the requisite properties.

149. What notations for the two operations were used by Boole ?

The sign + between two letters or symbols, as in $x + y$, Boole took to be the union of the subsets x and y.

The multiplication sign $\times$ was adopted by him to represent the intersection of sets.

The sign = represents the relationship of identity.

Boole selected his elements from a universal set or universe of discourse, the totality of which he designated by the symbol or number 1, and the complement of the set x by $1 - x$.

Notations have changed somewhat since Boole's day, so that union and intersection are usually denoted by $\cup$ and $\cap$ rather than by + and $\times$, and the symbol for the null set is ϕ rather than o, but the fundamental principles are those that were laid down by Boole about a century and half ago.

150. Which rules of the Boolean algebra are not valid in the ordinary Algebra ?

Some of the Boolean algebras based on union, intersection, complementation, and the relation of inclusion have unusual characteristics.

One is that this algebra has two distributive laws in stead of the usual one.

Unlike the algebra of most sets of numbers in which $x + x = 2x$ and $x.x = x^2$, the algebra of sets has $A \cup A = A$ and $A \cap A = A$, where union might correspond to + and intersection to ×.

Thus $A \cap A = A$,

which in Boole's notation reads as $A. A = A$ or $A^2 = A$, so that in Boole's algebra every element is square of itself.

Again in Boolean algebra, if $z x = z y$, where z is not the null set, it cannot be inferred that $x = y$.

Likewise, if $xy = 0$, in Boolean algebra it is not necessarily true that either x or y must be a null set.

Such assertions are not valid in ordinary algebra.

151. How is Boolean algebra a significant advancement toward abstraction ?

In Boolean algebra the primary interest is not in the elements of certain domains, but in the subsets of these domains and their inclusions, intersections and unions.

The nature of individual elements is not relevant.

Sets of elements rather than the elements themselves become the basic data, which turn out to be its distinctive feature leading to further abstraction.

Boole is therefore credited with having taken the first and the decisive step toward abstract algebra.

152. Where is Boolean Algebra applicable ?

Boolean Algebra has many practical applications in the physical sciences, in electric—circuit theory and particularly in the field of computers.

153. How about Cayley's Algebra ?

In Cayley's Algebra, multiplication is neither commutative nor associative, i.e., $ab \neq ba$, and $a(bc) \neq (ab)c$.

It appears that such a truncated algebra would have no physical significance, but remarkably enough it has been applied to the quantum theory.

154. How many algebras are there ?

There is no limit to the formation of algebras, because by adding or suppressing a few postulates a different algebra can always be framed.

Benjamin Pierce worked out multiplication tables for 162 algebras.

His son, C. S. pierce continued his father's work in this direction and showed that of all these algebras there are only three in which division is uniquely defined, namely, ordinary real algebra, the algebra of complex numbers, and the algebra of quaternions.

155. Are so many algebras in use ?

No, not at all !

Unless an algebra finds some practical applications, it's of no use. It constitutes only "a barren set of postulates and a sterile list of formulas" as remarked by a celebrated author.

In the early 19th century it was believed that there was only a single algebra, but the realization that an algebra was only an outcome of its postulates dispelled the belief as totally untrue.

156. Is then mathematics only a creation of the postulates and nothing beyond ?

That mathematics is postulational thinking in which valid conclusions are drawn from arbitrary premises has been recognized since the end of the 19th century, and has been the characteristic feature of mathematical activity of the 20th century.

Mathematical truths are, therefore, not absolute. They are only relative truths. A theorem is true if the postulates from which it is derived are true.

157. What then is the validity of the conclusions drawn in mathematics ?

A simple example is illustrative.

The statement

$$5 + 3 = 1$$

is obviously bad arithmetic.

But *if* we look upon the numerals as representing the days of the week, thus

Sunday – 1
Monday – 2
Tuesday – 3
Wednesday – 4
Thursday – 5
Friday – 6
Saturday – 7;

the numeral 5 stands for Thursday and three days after Thursday is always a Sunday, represented by 1.

The statement $5 + 3 = 1$ is no longer bad arithmetic.

Mathematical statements are, therefore, only relative truths and true only in the context for which they are meant.

The mathematical way of drawing conclusions depends upon such assertions as "*if* such a proposition is true of *anything, then* such and such another proposition is true".

Whether the first proposition is really true or what that *anything* is of which it is supposed to be true are matters of no concern at all.

158. What is the implication of the oft – quoted Russell's epigram of 1901 ?

Bertrand Russell worded the same idea in a witty and amusing way as follows :

"Mathematics may be defined as the subject in which we never know what we are talking about, nor whether what we are saying is true."

In spite of its humour, Russell's elucidation actually emphasizes the entirely abstract character of mathematics.

At first sight, Russell's epigram appears to degrade mathematics, but what is really intended is that in mathematics we deal with various entities without knowing or even caring to know their meaning.

Take for example the study of geometry, where we talk about entities such as points, lines, planes etc. Definitions like : a point is that which has position but no magnitude, or a line is the shortest distance between two points, may be good for a beginner but are not satisfactory.

One must not, therefore, feel concerned whether a point means an extremely small dot, or a line an extremely thin scratch, but merely specify in the beginning the axioms according to which the subject is to be developed. All the conclusions which follow logically are then derived.

Thus it is not required to know what exactly is meant by the basic terms or whether what is said about them is true, because if the axioms are true in a certain context, the assertions will also be true only in that context.

159. If mathematics is only abstraction, how does it yield correct results when applied to actual situations ?

The modern viewpoint is that one must distinguish between mathematics and *applications* of mathematics.

Really speaking, mathematics presents one model, and the situation where it is to be applied represents another so that the degree of success in application depends upon the degree of resemblance between the two models.

The first model must be ideal, i.e., abstract. To obtain good results, the second must resemble the first as close as possible.

160. How about the traditional conception of mathematics as the "science of magnitude and number" ?

Magnitude and number make an important part of the material to which mathematics has been applied, but they do not themselves constitute mathematics just as the canvas and the paints do not constitute the masterpiece an artist paints.

161. What is the modern view ?

The modern view is that if any topic is presented in such a way that it consists of symbols and the precise rules of operation upon these symbols, subject only to the requirement of inner consistency, then this topic constitutes mathematics.

The essential characteristic of mathematics is, therefore, not so much its content as its form.

In the mathematics of the 20th century the emphasis has been on abstraction, and there is an increasing concern with the analysis of broad patterns.

Thus mathematics is much more abstract than it had been traditionally supposed.

162. How is mathematics currently defined ?

Currently, mathematics is defined as :

"The postulational science in which necessary conclusions are drawn from specified premises".

The definition is, in essence, the same as the well-known definition of mathematics Benjamin Pierce gave in 1870 in connection with his work on algebras :

"Mathematics is the science which draws necessary conclusions."

163. Set again ! What is a set ?

The word 'set' means just what it says – any collection of objects or abstract entities constitutes a set.

The letters on this page make up a set. All odd numbers make up another.

164. How is a set specified ?

A set may be specified in one of the two ways.

The roster method or tabulation method lists all the elements in the set. The set of numbers between 3 and 8 can be specified as

$$X = \{4, 5, 6, 7\}.$$

A set may be denoted by some symbol, such as X above.

The second method is the descriptive method or what is called the set – builder form, which gives a rule for determining which things are in the desired set and which are not.

165. What operations can be performed on sets ?

There are three basic set operations : Union, Intersection and Complementation.

These have been explained earlier.

166. What is the set theory about ?

Set theory is the study of relationships existing among sets.

It underlies the language and concepts of modern mathematics – both pure and applied.

167. What is a subset ?

It has been explained earlier.

To recall :

A is called a subset of B if each element of A is an element of B. It is expressed as $A \subset B$.

If $A = \{1, 2,. 3\}$,

and $B = \{1, 2, 3, 4, 5\}$, then $A \subset B$.

The set of odd integers is a subset of the set of real numbers.

Also, each set is a subset of itself.

168. But how can a set be a subset of itself ?

It's the very definition of a subset that makes it possible.

In accordance with the definition, A will be a subset of A, if every element of A is an element of A, which is obviously true, so that A is a subset of A is a statement valid for every set A.

169. What is meant by the cardinality of a set ?

The cardinality of a set is the measure of the size of the set. For instance, all sets with exactly 7 elements have the same cardinality.

But this is not true for infinite sets. Not all infinite sets have the same cardinality.

170. What is an infinite set ?

A set is an infinite set if it has no last member.

An infinite set has the unique property that the whole is no greater than some of its parts.

For example, consider all the positive integers,

1, 2, 3, 4, 5, 6, 7, 8,..............,

and under each write its double, thus,

2, 4, 6, 8, 10, 12, 14, 16,..........

The numbers in the second row are exactly as many as those in the first row, because we got the second row by doubling the numbers in the first.

This leads us to conclude that there are just as many even numbers as there are whole numbers altogether.

The numbers 1, 4, 9, 16, 25, etc., are called square numbers, and the numbers 1, 8, 27, 64, 125, etc., are called cube numbers, as they are respectively the squares and cubes of the natural numbers. They are also as many as the whole numbers.

171. In what way does an infinite set differ from a finite set ?

An infinite set is similar to a proper part of itself, while a finite set is similar to no part of itself.

This constitutes the fundamental difference between the finite and the infinite set.

172. Is counting not necessary for the comparison of sets ?

No. One who does not know anything of counting is still able to compare two collections, and can determine which is the greater and which the less.

For example, it is easy to show that we have the same number of fingers on both the hands by simply matching finger with finger on each hand.

One – one correspondence or the matching process is the key to comparison.

Besides, since infinite sets cannot be counted as in the ordinary way, they can be compared only by the matching process as will be evident a little later.

173. What is counting ?

Counting is basically comparing.

Suppose we have a collection of 5 oranges, another of 5 apples, next of 5 pencils, yet another of 5 books, and so on. The only property common to them all is what could be termed as their 5 – ness. This is an abstract idea arising out of comparing.

The concept of number is therefore an abstraction. Numbers come much later in the scheme of counting. In the initial stage, the process of comparing works sufficiently well.

174. How is number defined ?

The concept of number depends upon the concept of class.

As before, consider a collection of 5 oranges, another of 5 apples, next of 5 pencils, yet another of 5 books, and so on. Such collections may be called classes.

The property which is possessed by all the classes similar to any of these classes is what is usually associated with the number 5.

Therefore the class of all classes which are similar to the class of oranges or the class of apples, etc., can be taken as the definition of the cardinal number 5.

This leads to the definition of number :

The number of a class is the class of all those classes that are similar to it.

175. Can two straight lines of unequal length have the same number of points on them ?

Yes. An easy construction makes it clear.

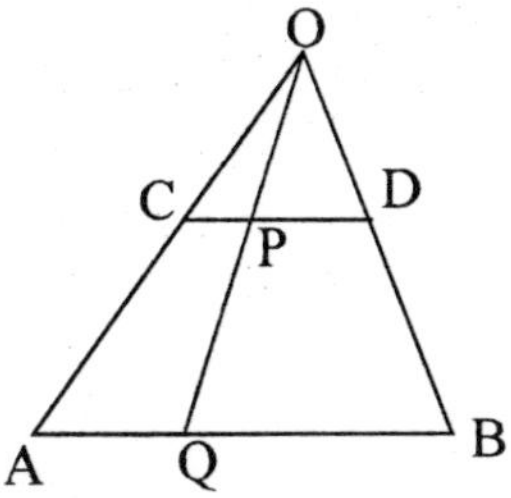

Take two segments A B and C D of different lengths. Place them parallel as in the figure, and let A C and B D be produced to meet at O.

Take any point, say Q, on A B, and join O Q.

Let O Q cut C D in P.

Then, to every point like Q on A B, there corresponds a point P on C D and vice versa.

The number of points on the two lines are therefore equal.

The conclusion appears to be contrary to common sense, but is true because the number of points in each segment is infinite.

Also, a line a millionth of an inch long has just as many points as a line stretching from the Earth to the furthest star.

176. Is the number of integers equal to the number of fractions ?

Common sense makes us believe that the fractions must be more numerous than the integers, because between any two integers there is an infinite number of fractions. But cantor discovered that the class of rational fractions is equivalent to the class of integers.

177. How is it proved ?

The complete set of rationals $\frac{p}{q}$ can be written down in a sequence by writing down first those rationals for which p + q = 2, then p + q =3, 4, 5, and so on, thus :

$\frac{1}{1};$

$\frac{1}{2}, \frac{2}{1};$

$\frac{1}{3}, \frac{2}{2}, \frac{3}{1};$

$\frac{1}{4}, \frac{2}{3}, \frac{3}{2}, \frac{4}{1};$

$\frac{1}{5}, \frac{2}{4}, \frac{3}{3}, \frac{4}{2}, \frac{5}{1};$

.... ;

....

Since each fraction may be written as a pair of integers, i.e., $\frac{3}{4}$ as (3, 4), a one – to – one correspondence is obtained between all the rationals and all the integers.

1	2	3	4	5	6	7	8	9	
↕	↕	↕	↕	↕	↕	↕	↕	↕	....
(1,1)	(2,1)	(1,2)	(1,3)	(2,2)	(3,1)	(4,1)	(3,2)	(2,3)	

It does not make any difference if from this sequence those rationals which are not in their lowest terms, e.g., $\frac{2}{2}, \frac{2}{4}, \frac{3}{3}, \frac{4}{2}, \ldots\ldots.$ are deleted.

Cantor called a set countably infinite if it could be put into one –to – one correspondence with the set of integers.

178. Is the number of points in a line also equal to the number of integers ?

No. The matching process reveals that the attempt to do so does not succeed in the case of real numbers.

For this it is enough to consider real numbers between 0 and 1.

If possible let the real numbers between 0 and 1 on any line A B be arranged in a sequence of the following type :

Point	
P_1	$.\underline{1}\,4\,2\,8\,5\,7\,1\,4\,2\,8\,5\,7.....$
P_2	$.3\,\underline{3}\,3\,3\,3\,3\,3\,3\,3\,3\,3\,3.....$

P_3	$.\,5\;0\;\underline{0}\;0\;0\;0\;0\;0\;0\;0\;0\;0.....$
P_4	$.\,6\;6\;6\;\underline{6}\;6\;6\;6\;6\;6\;6\;6\;6.....$
P_5	$.\,7\;2\;1\;8\;\underline{9}\;6\;5\;2\;4\;7\;1\;4.....$
P_6	$.\,8\;1\;2\;3\;5\;\underline{2}\;9\;1\;4\;1\;2\;3.....$
P_7	....
P_8	....
etc.	etc.

Here points are represented by their respective distances from one end A.

Also, for convenience, distances are represented by their corresponding non – terminating decimal equivalents.

Thus, $\frac{1}{7}$ is represented by. 142857142857......

$\frac{1}{2}$ " by. 500000000000....

$\frac{1}{3}$ " by. 333333333333....

Now, consider the number in the decimal notation formed by taking only the digits underlined in the above representation. This number is. 1 3 0 6 9 2.........

From this we can produce another number by changing each digit in some way. For instance, 1 may be replaced by the next number 2, 3 by 4, 0 by 1, 6 by 7, 9 by 0, 2 by 3, and so on.

Thus from the number. 130692......can be obtained another number. 241703......

This number is not included in the representation given above, because it differs in at least one place from each number listed above.

This leads us to conclude that all the real numbers between 0 and 1 cannot be arranged in a sequence, and that the matching process does not succeed in the case of real numbers.

179. Why are some infinite sets said to be countable and some others uncountable ?

An infinite set is said to be countably infinite if its elements can be put into one – to – one correspondence with the positive integers.

The set of positive integers and the set of rational numbers are both countable.

The set of real numbers is not countable because it cannot be put into one-to-one correspondence with the set of positive integers.

For the same reason, the set of irrational numbers is also uncountable.

Thus the infinity of the set of real numbers may be regarded as bigger than the infinity of the set of positive integers.

The infinity of the set of positive integers is denoted by the symbol a, and the infinity of the set of real numbers by c. The symbols a and c are called transfinite numbers.

180. What is meant by the Continuum Hypothesis ?

The conjecture that every infinite subset of the set of real numbers has the cardinal number either of the positive integers or of the entire set of real numbers is known as the Continuum Hypothesis.

This implies that no infinity of a power greater than a but less than c exists.

The conjecture was advanced by Cantor in 1878, but no one has yet produced an intermediate transfinite of this kind. It is believed that there is none such, although a strict mathematical demonstration has not yet come up.

This constitutes one of the unsolved problems of set theory.

181. What is meant by saying that the points on a line segment are "everywhere dense" ?

It means that between any two points on a line there is an infinite number of other points. The concept of two immediately adjoining points is, therefore, meaningless.

The class of real numbers and the class of points on a line segment, are both everywhere dense, and have the same cardinality, namely c.

Again, a line segment one—millionth of an inch long has as many points as there are in all three—dimensional space in the entire universe. There is no relation between the number of points on a line and its length.

182. What about the time continuum ?

Time is also "everywhere dense". Between any two instants of time an infinity of others may be interpolated. The time continuum has the same characteristics as the space continuum.

183. Does every description that seems to make sense define a set ?

No, not necessarily. Sometimes inconsistencies, such as the Barber's paradox arise.

184. What is meant by a paradox ?

By paradox is meant an argument in which it appears that an obvious untruth has been proved.

185. What is the Barber's paradox ?

The village barber shaves all those, and only those, who do not shave themselves. The question is : Does the barber shave himself ?

Either the barber shaves himself or he does not.

It he does, then he is shaving someone who shaves himself and therefore breaks his own rule. If he does not, then he breaks his rule by not shaving someone who does not shave himself.

This is called the Barber's paradox.

186. What is the way out of this paradox ?

It is clear that in either case there is a contradiction.

The following is the explanation :

The class of shavers is defined as the class of individuals not shaved by the barber.

Thus the class of shavers is defined by the barber, so that to include the barber himself in the word "all" is not legitimate and leads to an argument in a vicious circle.

187. What is the moral of this paradox ?

The moral is :

A set cannot be a member of itself.

Such paradoxes arise from the indiscriminate use of the word "all".

Here is another example.

Consider the following three statements :

1. This book has 856 pages.
2. The author of this book is Tagore.
3. The statements numbered 1, 2 and 3 are all false.

The question is : Is the statement 3 true or false ?

1 and 2 are false, but 3 is neither true nor false.

Indiscriminate use of the word "all" leads to similar logical inconsistencies as in the following examples.

A Cretan is made to say : "All Cretans are liars".

Here, if the statement is true, it makes the speaker a liar for telling the truth.

Another example: "All generalizations are false including this one".

188. Are paradoxes not mere tricky statements meant to emphasize the limitations of logic ?

Paradoxes appear to be tricky statements but they pose real problems in the foundations of logic, and have puzzled thinkers for centuries

It has to be conceded that if paradoxes can arise from apparently valid reasoning in set theory, they may arise anywhere in mathematics. Since generalisation is the soul of mathematics, it is necessary to examine whether in trying to extend a proposition from one specific instance to all situations, any contradiction has inadvertently crept in.

Logical paradoxes, therefore, play a role of the utmost importance in ascertaining whether mathematical advances are in order.

189. What is Russell's paradox ?

We have seen that indiscriminate use of the word "all" gives rise to some paradoxes.

In the same way, unrestricted freedom in the use of the concept of "set" also leads to contradictions.

Russell's paradox refers to an inconsistency which emerges out of such use :

Let S be the set of all sets, and only those sets, which do not contain themselves as an element. For instance, the set of apples belongs to S because the set of apples is not an apple.

Now consider the statement "S is an element of itself". Is this statement true or false ?

If it were true, S would be an element of itself, and therefore not in S by the definition of S. Thus the statement must be false.

If it were false, S is not an element of itself, and therefore is in S, by definition.

Thus in either case, a contradition is reached.

Someone had written "Logic is barren, whereas mathematics is the most prolific of mothers".

Russell's paradox prompted the celebrated French mathematician Poincare' to exclaim : "Logic is no longer barren ; she has brought forth a contradiction !"

190. How is the paradox removed ?

To remove the paradox Russell and Zermelo in 1908 suggested a way that is now generally accepted.

Sets are classified into types corresponding to the stages of their construction, such that the elements of a set must be at a lower level than the set itself. Collections such as a set of "all sets" are ruled out because, if such a set existed, it would not contain the sets of its own level and higher ones.

191. Where has set theory found applications ?

Set theory is of great value because of the clarity it brings to investigations concerning infinity.

Besides, it has applications to logic, computer science, and other branches of mathematics.

192. How is set theory received at the elementary level ?

It seems that the only reasonable way to present mathematics is to do it axiomatically. This brought about systematization of mathematics, and led to mathematical models with names like set, group, ring, linear space, topological space etc.

These are mathematical objects rather than objects that appear in everyday life.

But children and most college students think in very concrete terms and find it difficult to come to terms with such abstract objects. Their reaction to set theory, therefore, is not favorable.

It can be judged from the following anecdote :

A Russion peasant came to Moscow for the first time and went to see the sights. He went to the zoo and saw the giraffes.

"Look", he said, "at what the Bolsheviks have done to our horses".

That is what modern mathematics has done to simple geometry and simple arithmetic.

❖❖❖

3

Analysis and Analyses

1. What are the basic fields of mathematics ?

Algebra, Geometry and Analysis.

2. What accounts for such a division ?

All mathematics arises out of problem solving, and a single method or approach cannot suffice for all the situations.

These three provide fairly distinct methods of approach. They are, therefore, termed as the basic fields of mathematics.

3. How did the word Analysis come to stay ?

Calculus originated in the 17th century and in its investigations utilized the analytic method of approach so that those branches of mathematics whose roots are in calculus are often known by the collective name of Analysis.

Before the advent of calculus it was geometry which supplied the main fabric of mathematical thought, and geometry uses the method of construction or what is called the synthetic method.

4. What does analysis deal with ?

Analysis is the study of functions, which means study of relations among continuously changing quantities.

It is totally different from elementary mathematics which deals with constant magnitudes.

By the 15th and 16th centuries scientists had become more and more concerned with problems based on varying quantities and their relations. Some of these problems could not be solved by the methods of arithmetic and algebra. Analysis provided the necessary tool, and advanced not only science and mathematics but also proved to be the master key to the entire technological progress in the centuries following Newton.

5. What has been the motivating force for Analysis ?

Mechanics, Physics, and technology, in particular, have provided the motivating force for all branches of analysis.

Everywhere around us Nature seems to be in a state of constant flux. From the smallest particles to the most massive stars, there is motion, flow and change. Broadly speaking, therefore, every natural phenomenon is the domain of Analysis.

6. As a mathematical tool, how is Analysis different from Algebra and Geometry ?

For this consider the motion of a falling body.

A freely falling body covers a distance equal to 16 feet in the first second of its motion. In 2 seconds the total distance covered is 64 feet; in 3 seconds, 144 feet ; in 4 seconds, 256 feet ; in 5 seconds, 400 feet; and so on.

The following table gives the correspondence between time (t) in seconds and the distance (s) fallen in feet :-

Time (t) in seconds	:	0,	1,	2,	3,	4,	5, ...
Distance (s) fallen in feet	:	0,	16,	64,	144,	256,	400, ...

Clearly the rate at which the body falls is not uniform during the motion. In other words, the speed itself is increasing every instant.

What is the speed at any particular *instant* and how can we calculate it ?

It may be recalled that speed is ordinarily calculated by dividing distance covered by the time taken to do so. But that determines only the *average speed* during the given interval of time, while what is being looked for is the speed at any given instant of time and not during any interval of time howsoever small.

This apparently simple problem cannot be tackled by the methods of Algebra and Geometry. Only Analysis provides the powerful tool.

7. Why is this example so important ?

Because it is representative of a fundamental problem : the problem of measuring rate of change, i.e., the rate at which one variable quantity changes with regard to another.

Most scientific studies are about varying quantities and their mutual relations, and it is impossible to interpret such relations without the basic concepts of analysis.

In this problem, time is one varying quantity and distance the other. As time increases, distance also increases but not uniformly, so that the speed itself is at no time constant.

8. How does analysis solve this problem ?

In this problem we observe that :

$$16 = 16 \times 1^2,$$
$$64 = 16 \times 2^2,$$
$$144 = 16 \times 3^2,$$
$$256 = 16 \times 4^2,$$
$$400 = 16 \times 5^2, \text{ and so on.}$$

This leads us to the relationship between s and t as

$$s = 16\,t^2 \qquad ....(1)$$

We consider what will happen to s if we make an extremely small change in t.

A mathematician represents a small bit by the letter "d". This means "an extremely small amount of". Thus ds (pronounced dee-ess) represents an extremely small distance s. Similarly dt (pronounced dee-tee) means an extremely small amount of time t.

To t we add a little bit more of t, and we obtain t + dt. This produces a small change ds in s, which becomes s + ds.

The equation (1), therefore, takes the form :

$$s + ds = 16\,(t + dt)^2$$

or $$s + ds = 16t^2 + 32t\,(dt) + 16\,(dt)^2 \qquad .....(2)$$

Subtracting (1) from (2), we get $ds = 32\,t\,(dt) + 16\,(dt)^2$

Now, the last term on the right side is extremely small, so small in value that it may be ignored entirely. We then have

$$ds = 32\,t\,(dt)$$

Since instantaneous velocity is $\frac{ds}{dt}$, we have

$$\frac{ds}{dt} = 32\,t \qquad .....(3)$$

This equation tells us that the velocity of a falling body at any instant of time t is equal to 32 t.

For instance, velocity attained just at the end of 3 seconds is 32 x 3 = 96 feet per second.

Similarly, velocity attained just at the end of 3.1 seconds is 32 x 3.1 = 99.2 feet per second.

9. What are the basic concepts of Analysis ?

The basic concepts of Analysis are : Variable, Function, and Limit.

10. What is Variable ?

If some entity changes in value, it may be called a variable.

Examples of variable are time, distance, area, volume, height, velocity, angle of rotation, etc.

Given two variables, say distance and time, such that the first depends on the second, we call the first the dependent variable and the second the independent variable.

Likewise, since area of a circle depends on its radius, we call area the dependent variable and radius the independent variable.

11. What is Function ?

To say that y is a function of x means, in mathematics, only that to each possible value of x there corresponds a definite value of y.

Functions are exrpressed by equations, graphs, tables of values or verbal sentences.

That distance s is a function of time t is expressed symbolically as :

$$s = f(t), \qquad .....(1)$$

where f is only a convenient short form for "function of" or "depends on".

It is spoken as : s is a function of t.

Likewise, that area of a circle, is a function of its radius r, is expressed as :

$$A = f(r) \qquad .....(2)$$

This is spoken as : A is a function of r.

Like (1) and (2) there could be several other equations covering different situations.

All such equations can be generalised into one equation and expressed as :

$$y = f(x) \qquad .....(3)$$

where y stands for any dependent variable, x for any independent variable, and f for any rule connecting y and x.

This is the most important and the most frequently occurring equation in Analysis.

12. What makes this equation so important ?

The power and importance of this equation lies in the complete abstraction of its form in the sense that it does not concen itself with any particular variable or function but with variables and functions in general.

Study of functions is the subject matter of Analysis. But, here, in stead of studying particular functions, a general study of functions is done.

A generalised study of this kind opens up a vast scope of applications because one formula or one theorem in such study covers an entire infinity of all particular cases.

13. What is limit ?

Two kinds of limit are of particular interest in the calculus.

One is the limit of a quotient, used in the Differential Calculus.

The other is the limit of a sum, which forms the basis of the subject matter of the Integral Calculus.

However, there is one more important limit : the limit of an infinite sequence, or the related concept of the limit of the sum of an infinite series.

The concept of limit is of fundamental importance to many domains of Analysis and plays a central role in the Complex Analysis and the Fourier Analysis in particular.

14. What is meant by the limit of a function ?

Strictly speaking, all limits involve functions.

Thus a limit is a number that the function approximates closely when the independent variable is subject to some restriction.

For example, the limit of $\frac{1}{x}$ is zero, if x increases beyond all bounds, taking positive values.

For instance,

if $x = 100$, $\frac{1}{x}$ equals .01,

if $x = 10000$, $\frac{1}{x}$ equals .0001,

if $x = 1000000$, $\frac{1}{x}$ equals .000001,

and so on.

By letting x take up sufficiently large values, $\frac{1}{x}$ assumes extremely small values almost indistinguishable from zero, but no value, however large will make it actually zero. This is what is meant by saying that the limit of $\frac{1}{x}$ is zero.

15. What is Heine's $\in - \delta$ definition of the limit of a function ? What are $\in$ and δ ?

$\in$ and δ are letters from the greek alphabet and represent small quantities.

Let f be the function, then Heine's definition runs as follows :

A function f has the limit k as x approaches a if, for every positive number $\in$, there is a number δ such that $|f(x) - k| < \in$, if $0 < |x - a| < \delta$.

The basic idea is the same but has been expressed in a form that conforms well with mathematical precision.

Note : $|x - a|$ is called the modulus of $x - a$, and means the positive value of $x - a$, i.e.,

if $x > a$, $|x - a| = x - a$,

and if $a > x$, $|x - a| = a - x$.

16. What is remarkable about this definition ?

The definition is the result of more than a hundered years of trial and error, and provides in very few words a sound mathematical basis to an important concept.

It utilizes only the most commonly used notions like number, the operation of addition (or its inverse, subtraction), and the relationship "less than", and above all does not make use of the concept of a variable so indispensable in calculus.

17. What is meant by the limit of a quotient ?

Differential calculus arises from the study of the limit of a quotient, $\frac{\Delta y}{\Delta x}$, as the denominator Δx approaches zero,

where y and x are variables,

y is some function of x, and

Δy and Δx represent extremely small changes in y and x respectively.

As Δx, the small change in x, tends to zero; Δy, the corresponding small change in y, also tends to zero, but the ratio tends to a definite quantity, called its limit.

The limit is denoted by $\frac{dy}{dx}$ and is called the derivative of y with resepct of x. It represents the rate of change of y with respect to x.

The process of determining derivatives is called differentiation.

Derivatives of all commonly used functions can be found by this method but the process is facilitated by formulas for derivatives of all common functions.

A few of the formulas are as follows :-

If $y = x^n$, $\frac{dy}{dx} = n\,x^{n-1}$,

If $y = \sin x$, $\frac{dy}{dx} = \cos x$,

If $y = \cos x$, $\frac{dy}{dx} = -\sin x$,

If $y = \log_e x$, $\frac{dy}{dx} = \frac{1}{x}$,

If $y = e^x$, $\frac{dy}{dx} = e^x$

Also, the derivative of a constant is zero.

18. What is meant by the limit of a sum ?

This kind of limit occurs in the problem of finding the area under a curve.

Since every equation represents a curve and every curve has an equation, the graph of a function may be viewed as the geometrical representation of the points whose coordinates satisfy the equation y = f (x).

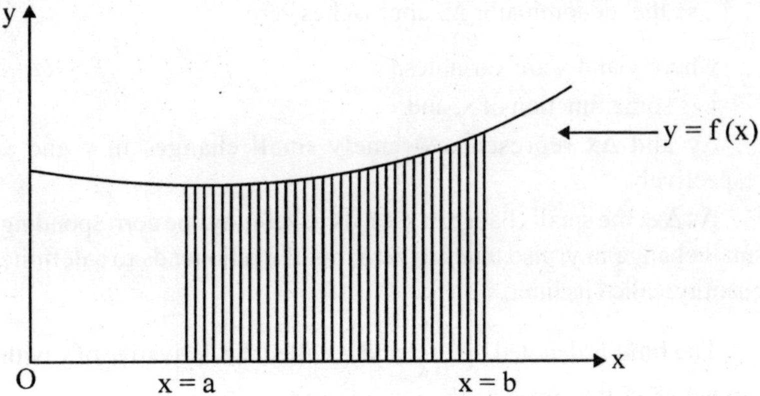

Now consider the problem of determining the area under a given curve y = f (x) between two ordinates, say x = a and x = b, and above the axis of x.

The interval between a and b is divided into n equal divisions, and the area cut into n strips, where n is any number like 10, 20, 50, 100 etc.

Supposing the strips to be nearly rectangular, if the areas of all these strips be added together, the sum will be approximately equal to the required area.

This approximation can be improved by increasing the number of strips and thereby decreasing their size.

In the limit when n approaches infinity, or which is the same as saying that the width of each strip approaches zero, the sum is equal to the area under the curve between the vertical lines at x = a and x = b.

This limit is denoted by $\int_a^b f(x)\,dx$, and is called the integral of f(x) with respect to x, the integral being taken from a to b, a and b are called the lower and the upper limit respectively.

The process is called integration, which means summing up small bits to get the whole.

19. What is infinite sequence, infinite series and the limit of its sum ?

A set of numbers ordered according to some rule and does not have a last term is said to be in infinite sequence.

The simplest infinite sequence is the set of integers :

1, 2, 3, 4, ...

the dots at the end mean that there is no last term.

The following is another infinite sequence :

$$1, \frac{1}{2}, \frac{1}{4}, \frac{1}{8}, \frac{1}{16}, \frac{1}{32}, ...$$

Infinite series is the addition of the terms of the infinite sequence.

The following is the corresponding infinite series :

$$1 + \frac{1}{2} + \frac{1}{4} + \frac{1}{8} + \frac{1}{16} + \frac{1}{32} + ...$$

The sum of the first 5 terms of this series is 1.9375, the sum of the first 10 terms of the series is 1.9980 (nearly), the sum of the first 15 terms of the series is 1.9997 (nearly).

What is interesting to note is that no matter how many terms of this series are taken, the sum cannot exceed a certin bound called its limit.

The limit of the sum in this case is 2, and the series said to converge to the limit 2. Such a series is called a convergent series.

20. Does every series converge to a limit ?

No. Consider the following series :

$$1 + \frac{1}{2} + \frac{1}{3} + \frac{1}{4} + \frac{1}{5} + \frac{1}{6} + ...$$

The sum of this series has no bound or limit.

It a sufficient number of terms of this series is taken, the sum can be made bigger than a billion billion billion.

Seems incredible but is true !

It can be demonstrated as under :

The 3rd and 4th terms are $\frac{1}{3}$ and $\frac{1}{4}$. They are together greater than $\frac{1}{4}+\frac{1}{4}$, i.e., $\frac{1}{2}$

The next four terms are $\frac{1}{5}, \frac{1}{6}, \frac{1}{7}$ and $\frac{1}{8}$. They are together greater than $\frac{1}{8}+\frac{1}{8}+\frac{1}{8}+\frac{1}{8}$ or $\frac{4}{8}$ or $\frac{1}{2}$.

The next eight terms are $\frac{1}{9}, \frac{1}{10}, \frac{1}{11}, \frac{1}{12}, \frac{1}{13}, \frac{1}{14}, \frac{1}{15}$ and $\frac{1}{16}$.

They are together greater than

$\frac{1}{16}+\frac{1}{16}+\frac{1}{16}+\frac{1}{16}+\frac{1}{16}+\frac{1}{16}+\frac{1}{16}+\frac{1}{16}$ or $\frac{8}{16}$ or $\frac{1}{2}$; and so on.

The series is, therefore greater than

$$1+\frac{1}{2}+\frac{1}{2}+\frac{1}{2}+\frac{1}{2}+\ldots\ldots$$

By taking sufficient number of terms of this series, its sum can be made to exceed any number howsoever large.

Such a series is said to be divergent.

But the following series is convergent :

$$1-\frac{1}{2}+\frac{1}{3}-\frac{1}{4}+\frac{1}{5}-\frac{1}{6}+\ldots..$$

the limit of its sum is $\log_e 2$ or .6931 (nearly)

However, it is not equal to

$$\left(1+\frac{1}{3}+\frac{1}{5}+\ldots\ldots\right)-\left(\frac{1}{2}+\frac{1}{4}+\frac{1}{6}+\ldots\ldots.\right),$$

since the series in each bracket is divergent, and it means nothing to subtract the second from the first.

21. What are Arithmetic, Geometric and Harmonic Progressions ?

Arithmetic progression :

An arithmetic progression is a sequence in which each term is obtained from the previous term by adding to it a given number d, called the common difference.

It has the general form :

a , a + d, a + 2 d, ..., a + (n −1) d, ...,

where a is the first term, and a + (n − 1) d is the nth or the general term.

The progression 3, 7, 11, 15, ... is an arithmetic progression with a = 3 and d = 4.

The sum, S, of the first n terms is given by the formula :

$$S = \frac{n}{2}\{2a + (n-1)d\}$$

Accordingly, the sum of the first 10 terms of the progression 3, 7, 11, 15, ... will be

$$\frac{10}{2}\{2 \times 3 + (10-1) \times 4\}, \text{ which reduces to } 210.$$

Geometric Progression :

A geometric progression is one in which each term is obtained by multiplying the previous term by a given number r, called the common ratio.

It has the general form :

$a, ar, ar^2, ...ar^{n-1}, ...$

where a is the first term, and ar^{n-1} is the nth or the general term.

The progression 1, 2, 4, 8, 16,......... is a geometric progression with a = 1 and r = 2.

The sum S, of the first n terms of the geometric progression is given by the formula :

$$S = \frac{a(r^n - 1)}{r - 1}.$$

Accordingly, the sum of the first 10 terms of the progression 1, 2, 4, 8, ... will be

$$\frac{1 \times (2^{10} - 1)}{2 - 1} \text{ or } 2^{10} - 1 \text{ or } 1023.$$

Harmonic progression :

A harmonic progression is one in which the terms are the reciprocals of an arithmetic progression.

The progresion 1, $\frac{1}{2}$, $\frac{1}{3}$, $\frac{1}{4}$, ... is a harmonic progression.

This type of progression has no general formula to express its sum.

22. How has the term harmonic come to stay ?

Harmonic is a physical term describing the vibration in segments of a sound-producing body, and is intimately connected with the harmonic progression.

When a string vibrates in its whole length, it also vibrates simultaneously in segments of halves, thirds, fourths, etc. These segments form what in algebra is known as a harmonic progresion, and each segment vibrates respectively twice, three times, four times, etc., as fast as the whole string.

The vibration of the whole string produces the fundamental tone, and the segments produce weaker subsidiary tones.

23. What are the applications of the infinite series ?

Infinite series may be used to determine the value of pi (π), a constant that appears in many branches of science.

The symbol π is used to denote the ratio of the circumference of a circle to its diameter.

To compute its value to any number of decimal places the following series may be used :

$$\frac{\pi}{4} = 1 - \frac{1}{3} + \frac{1}{5} - \frac{1}{7} + \frac{1}{9} - \frac{1}{11} + ...$$

The value of π, to five decimal places, is 3.14159.

Several other infinite series are also used to compute the value of π.

The letter e likewise occurs frequently in mathematics, for example, as a base in logarithms and in compound interest formula.

Its value to any number of decimal places may be found from the infinite convergent series :

$$e = 1 + \frac{1}{\underline{|1}} + \frac{1}{\underline{|2}} + \frac{1}{\underline{|3}} + \frac{1}{\underline{|4}} + ...$$

The sign $\underline{|\;\;}$ is known as the factorial sign. Thus $\underline{|3}$ means $1 \times 2 \times 3$, or 6 ; and $\underline{|5}$ is $1 \times 2 \times 3 \times 4 \times 5$, or 120.

Like π , e is also non-ending.

The value of e, to five decimal places, is 2.71828.

Infinite series are also used in computing trigonometrical tables. The sines of various angles can be computed by the following series:

$$\sin x = x - \frac{x^3}{\underline{|3}} + \frac{x^5}{\underline{|5}} - \frac{x^7}{\underline{|7}} + \ldots\ldots$$

Here x is in radian measure so that if the angle is equal to 30°, $\frac{\pi}{6}$ should be substituted for x in the series.

$$\text{Thus } \sin 30° = \sin\frac{\pi}{6}, \quad \because 180° = \pi \text{ radians.}$$

$$= \frac{\pi}{6} - \frac{\left(\frac{\pi}{6}\right)^3}{6} + \ldots\ldots\ldots$$

$$= \frac{3.14}{6} - \frac{(3.14)^3}{216} + \ldots.$$

$$= .52 - .02$$

$$= .5.$$

Similar series exist for cosine and tangent.

Likewise, logarithm tables can be computed to any number of decimal places by the use of infinite series.

24. What is zeno's paradox about Achilles and the Tortoise ?

A tortoise has a head start on Achilles. Zeno's paradox lies in arguing that although Achilles runs faster than the tortoise, he would never catch up with the tortoise. His argument is as follows :

Suppose A, B, C, D are some points on a straight track with Achilles at A and the tortoise at B.

While Achilles goes from A to B, the tortoise goes from B to C, and while Achilles goes from B to C, the tortoise goes from C to D, and so on, the process never ending, and Achilles always remaining behind the tortoise.

The explanation of the fallacy lies in the fact that motion is measured by intervals of distance covered in unit time and not by the number of points in any interval.

25. But what has it to do with the infinite series ?

A slight variation of zeno's argument makes it clear.

Achilles must first cover half the distance between him and the tortoise, then half of the remaining distance, then again half of what remains, and so on. It follows that some portion of the distance will always remain uncovered.

The successive distances to be covered form an infinite geometric series :

$$\frac{1}{2} + \frac{1}{4} + \frac{1}{8} + \frac{1}{16} + \ldots\ldots$$

each term of which is half of the one before.

Zeno assumed that since the terms of this series are infinite in number, their sum must also be infinite. And an infinite distance cannot be covered by any runner. So Achilles would not be able to cover even the intervening distance, much less overtake the tortoise.

In Zeno's time (4th century B. C.) an infinite series was believed to be always divergent, but today we know that this series is convergent and that the limit of its sum is 1, and not infinity.

26. What is Fibonacci Sequence ?

The following is an example of such sequence :

1, 1, 2, 3, 5, 8, 13, 21, ...

where each term after the first two is the sum of the two terms immediately preceding it.

Thus $2 = 1 + 1, 3 = 2 + 1, 5 = 3 + 2, 8 = 5 + 3, 13 = 8 + 5$, and so on.

This sequence occurs in connexion with the answer to the following question :

How many pairs of rabbits will be produced in a year, beginning with a single pair, if in every month each pair bears a new pair which becomes productive the second month on ?

A property of the sequence is that no two adjacent terms have a common factor.

27. What is hypergeometric series ?

The following is known as the hypergeometric series :

$$1+\frac{\alpha\,.\,\beta}{1\,.\,\gamma}\,x+\frac{\alpha\,(\alpha+1)\,.\,\beta\,(\beta+1)}{1\,.\,2\,.\,\gamma\,(\gamma+1)}\,x^2+\,......$$

It is usually denoted by F (α, β, γ, x).

28. Why is it called hypergeometric ?

In the above series, it we put $\alpha = 1$, and $\beta = \gamma$, the series reduces to the form

$$1 + x + x^2 + x^3 + \,.....,$$

which is a geometric series.

Since the series reduces to a geometric series as a particular case, it is called hypergeometric.

29. How is Analysis different from calculus ?

Analysis is the name given to the branches of mathematics that use the concept of limit.

Since calculus uses the concept of limit, it is considered a branch of analysis.

But the spirit of analysis is to treat and develop the subject matter of calculus by methods that make no appeal to geometrical concepts.

Current research in Analysis is mostly on a level of development far beyond the elements of calculus. The ideas of calculus, however, persist though in much more generalized and abstract form. Theory of functions and Functional Analysis are some such domains.

30. What does calculus basically deal with ?

Calculus commonly means the union of the Differential Calculus and the Integral Calculus.

The Differential Calculus deals with rates or derivatives.

The Integral Calculus has two aspects.

One is to view integration as the inverse of differentiation; that is, given a derivative, to determine the function which produces that derivative. Such a function is called an inverse derivative or an integral.

The other is to compute lengths of curves, areas bounded by curves, volumes bounded by surfaces and such other problems.

The two calculuses supplement each other and together they assist in understanding the eternal flux of nature.

31. What sort of problems are studied in the Differential Calculus ?

In the Differential Calculus are studied problems involving continuously changing quantities.

Thus it deals with all problems related to velocity, acceleration, rate of flow of heat or fluids, rates of chemical reactions, slopes of curves, etc.

Calculus enters wherever varying quantities and their relations are under study. Thus problems like the following and several such others naturally come into its domain :

(i) Grape juice is being fermented to wine. How fast is the grape juice being converted to alcohol ?

(ii) A rocket is fired directly upward with a veocity of 10 kilometres per second. What will be its velocity say 20 seconds after launching ? What is the highest altitude to which it will rise ? What time will it take to reach that altitude ? How long will the rocket's entire flight take ?

Since differential calculus involves derivatives, it also deals with such topics as maxima and minima, and rates of change of functions.

32. Who invented the calculus ?

Isaac Newton (1642-1727) and Gottfried Liebnitz (1646-1716) are credited with the discovery of calculus, thought it grew out of the efforts of many mathematicians to solve two geometric problems which interested the Greeks. One was the problem of finding the tangent to a curve at a given point, and the other that of determining the area bounded by a curve.

Calculus is one of the great triumphs of the human intellect, and next to Euclidean geometry is the greatest creation in all of mathematics.

33. What precisely does the term derivative imply ?

The derivative has two important interpretations :

(1) As the slope of a curve,

(2) As the velocity of a moving particle.

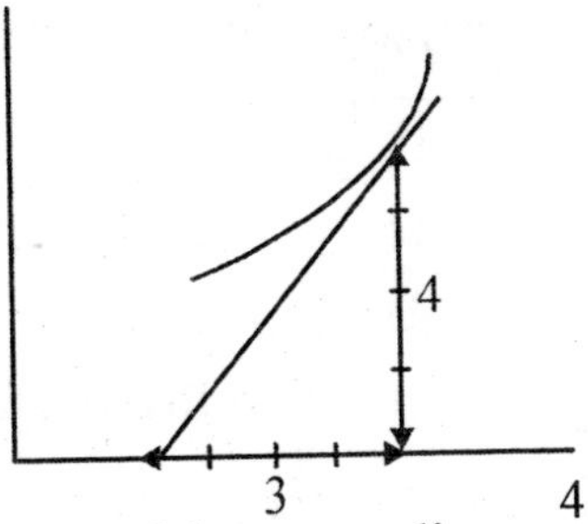

Slope of the tangent line = $\frac{4}{3}$

Slope of a curve at a point is measured by the slope of the tangent line to the curve at that point.

The value of the derivative at the point is equal to the slope there.

From the point of view of mechanics, the derivative represents the velocity of motion at the given instant of time or the instantaneous velocity.

34. How are the slope and the velocity related ?

It turns out that the problem of determining velocity is equivalent to that of drawing a tangent to the curve and determining its slope.

Mathematically, if s = f (t) gives the rule how distance s varies with time, then velocity = $\frac{dS}{dt}$.

If the graph of the curve s = f (t) be drawn, the slope of the tangent line at any point will be equal to the value of $\frac{dS}{dt}$ at that point.

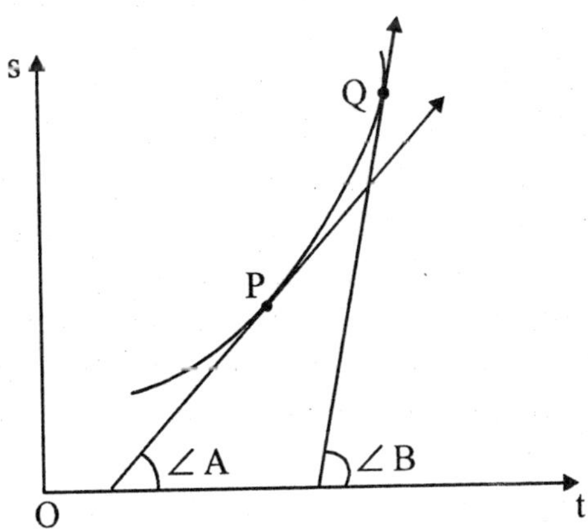

$$\text{Velocity at P} = \left[\frac{dS}{dt}\right]_{\text{at P}} = \tan A,$$

$$\text{Velocity at Q} = \left[\frac{dS}{dt}\right]_{\text{at Q}} = \tan B$$

35. How is the sign of the derivative related to the nature of the function ?

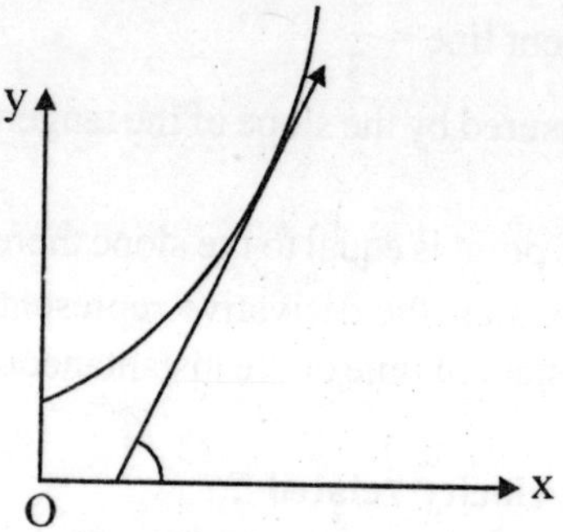

Slope angle less than a right angle
Increasing Function

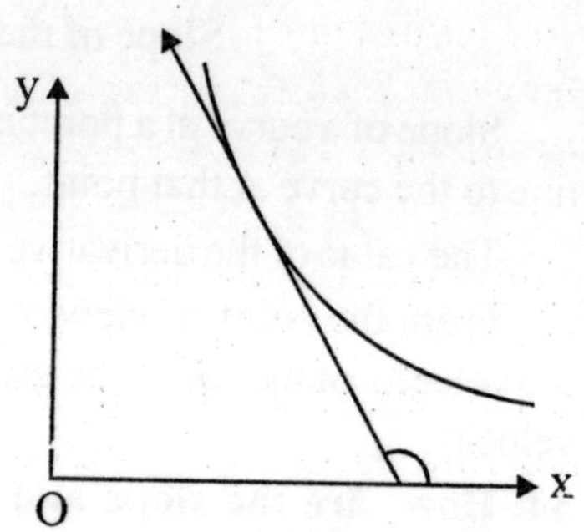

Slope angle greater than a right angle
Decreasing Function

A function is increasing at points where the slope angle is less than a right angle. This corresponds to a positive value of the derivative.

A function is said to be increasing if y also increases when x increases.

A function is decreasing at points where the slope angle is greater than a right angle. This corresponds to a negative value of the derivative.

A function is said do be decreasing if y decreases when x increases.

36. What happens when the derivative vanishes ?

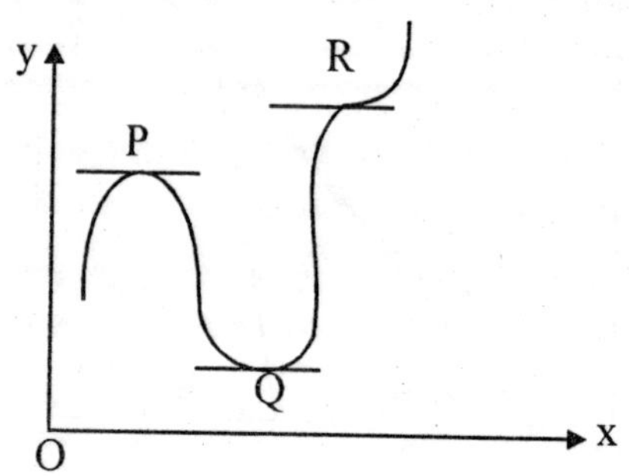

If the tangent line to the curve at a point is parallel to the axis of x, the slope there is zero, and the value of the derivative is also zero.

The function at such a point is neither increasing nor decreasing, but it might normally attain a value which is either a maximum or a minimum.

In the figure above :

At P, the function has a maximum,

At Q, the function has a minimum,

At R, the function has neither a maximum nor a minimum nor a minimum. Such a point is called a point of inflexion. At such a point the tangent crosses the curve.

37. How is the maximum or minimum of a function determined without drawing its graph ?

If the equation of the first derivative is set equal to zero, and solved, then it gives the values of x for which the function attains a maximum or a minimum.

For example, if $y = x^3 - 6x^2 + 9x$, ...(1)

then $\frac{dy}{dx} = 3x^2 - 12x + 9$.

Putting this equal to zero, and solving we get ;

$3x^2 - 12x + 9 = 0$,

or $3(x^2 - 4x + 3) = 0$,

or $x^2 - 4x + 3 = 0$, on dividing both the sides by 3

or $(x - 1)(x - 3) = 0$,

or $x = 1, 3$.

Thus the function has extreme points corresponding to x = 1 and x = 3. Extreme points is the common name for points having maximum or minimum.

Putting x = 1 in equation (1), we get y = 4, the maximum value.

Putting x = 3, we get y = 0, the minimum value.

38. What is the second derivative ?

The derivative of the derivative is called the second derivative and is denoted by $\frac{d^2y}{dx^2}$

Thus, if $y = x^4$, then $\frac{dy}{dx} = 4x^3$,

and $\frac{d^2y}{dx^2} = \frac{d}{dx}\left(\frac{dy}{dx}\right) = \frac{d}{dx}(4x^3) = 4 \times 3x^2 = 12x^2$.

The process can be continued to obtain derivatives of still higher order, i.e., $\frac{d^3y}{dx^3}$, $\frac{d^4y}{dx^4}$, etc.

39. What does the sign of the second derivative imply ?

Just as the sign of the first derivative determines whether the function is increasing or decreasing, so the sign of the second derivative determines the side toward which the graph of the function will be curved.

The positive sign of the second derivative indicates "convex downward" curve, and the negative sign "convex upward" curve.

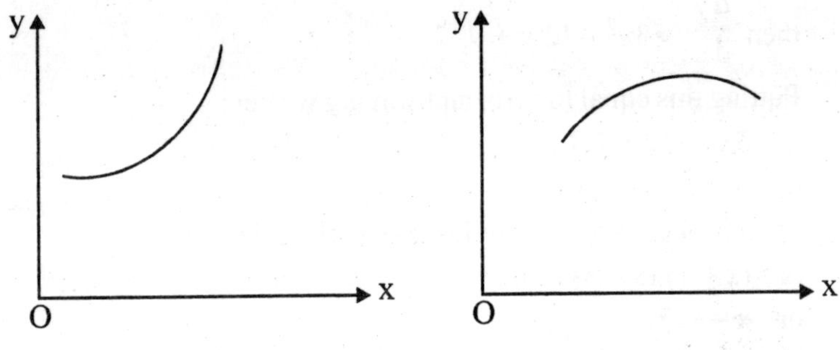

Second derivative positive convex downward

Second derivative negative convex upward

Whether an extreme point is a maximum or a minimum is also determined by the sign of the second derivative.

40. How is it determined ?

The positive sign of the second derivative indicates a minimum while the negative sign a maximum.

From the pervious example we have

$$y = x^3 - 6x^2 + 9x,$$

$$\therefore \frac{dy}{dx} = 3x^2 - 12x + 9,$$

$$\text{and } \frac{d^2y}{dx^2} = 6x - 12.$$

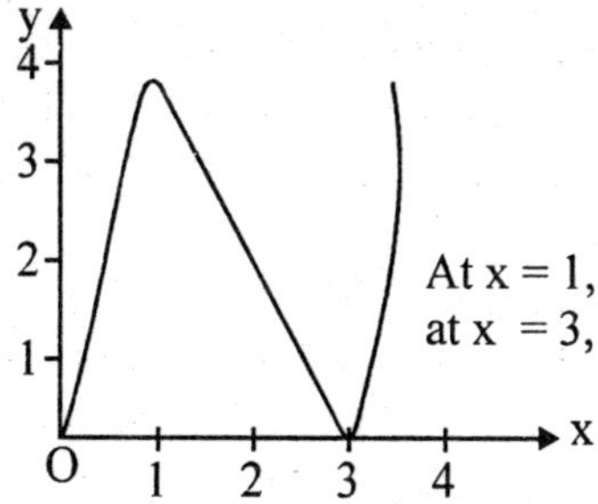

At x = 1, y is max.
at x = 3, y is min.

Putting $\frac{dy}{dx}$ equal to zero, i.e., $3x^2 - 12x + 9 = 0$, we get as before $x = 1$ and $x = 3$.

When $x = 1$, the value of $\frac{d^2y}{dx^2}$ equals $6 \times 1 - 12 = -6$, which is negative. Therefore, $x = 1$ corresponds to the maximum value.

When $x = 3$, the value of $\frac{d^2y}{dx^2}$ equals $6 \times 3 - 12 = 6$, which is positive. Therefore, $x = 3$ corresponds to the minimum value.

41. What are everyday applications of maxima and minima ?

The following are some interesting results having everyday applications.

1. Of all the rectangles with given perimeter, the square has the maximum area.

In other words, of all the rectangles with given area, the square has the least perimeter.

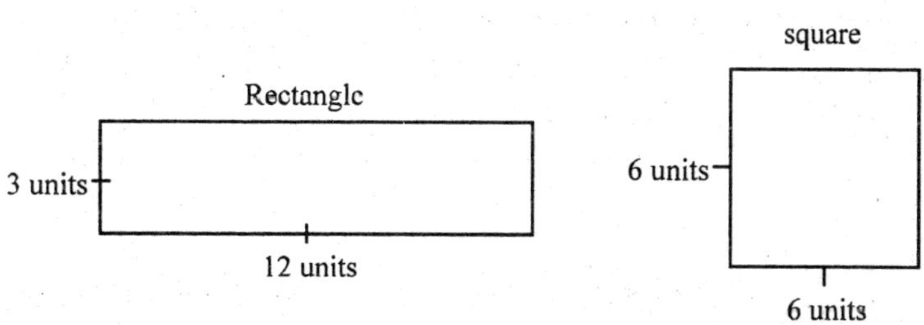

Area = 36 sq. units
Perimeter = 30 units

Area = 36 sq. units
Perimeter = 24 units

2. For a closed right circular cylinder of maximum volume and given surface area, the height is equal to its diameter.

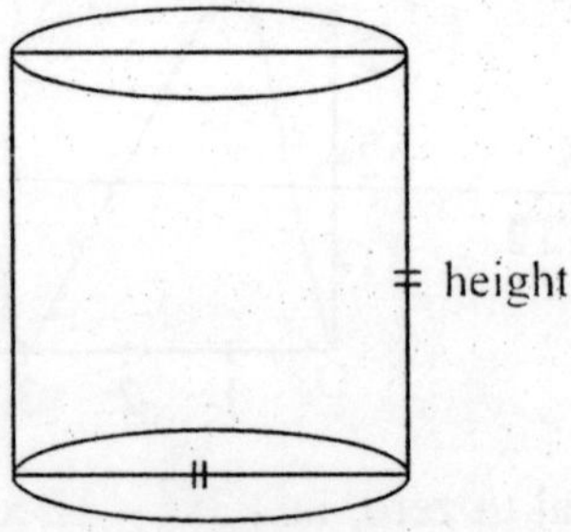

In other words, for a closed right circular cylinder of given volume and least surface, the height is equal to its diameter.

3. For an open cylinder of maximum volume and given surface, the height is equal to the radius of the base.

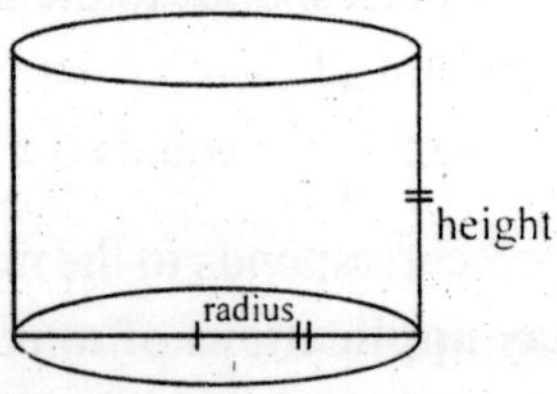

Or, for an open cylinder of given volume and least surface, the height is equal to the radius of the base.

4. The rectangle of maximum area that can be inscribed in a circle is a square.
5. Of all geometrical figures the circle encloses maximum area for given perimeter or that the circle has the least perimeter for given area.
6. For given perimeter, a triangle has maximum area when it has all sides equal.
7. Of all solids, the sphere has maximum volume for given surface area, or that the sphere has minimum surface area for given volume.

42. What is the significance of the second derivative in mechanics ?

The second derivative is of great physical importance in mechanics. Since the first derivative represents the rate of change of displacement, i.e., the velocity of a body, the second derivative represents the rate of change of velocity or more simply the acceleration of the body at time t.

Suppose that the distance s that a body falls in time t seconds is given by : $s = 16\,t^2$.

Then the velocity is given by $\frac{ds}{dt} = 32\,t$, and the acceleration by $\frac{d^2s}{dt^2} = 32$.

This equation tells us that all bodies fall at the same constant rate of acceleration (that is, if we neglect the resistance due to air).

43. Why are forces expressed in terms of second derivatives ?

Since forces are proportional to the accelerations they induce, forces essentially introduce second derivatives.

44. What is Heron's problem of the light ray ?

The problem is one of determining minimum distance. It is as follows :

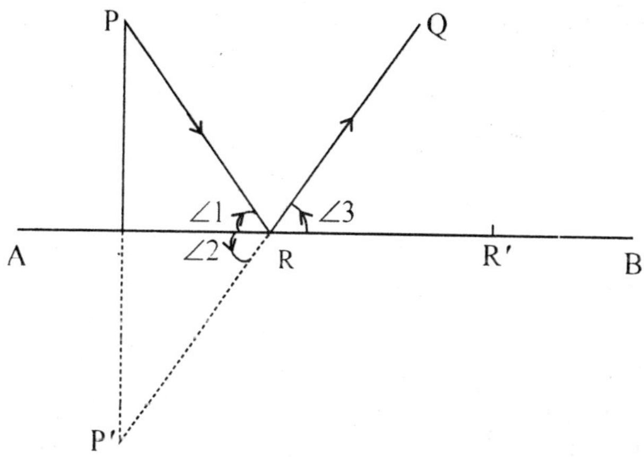

P and Q are two points on the same side of a reflecting surface AB. For what point R on AB is PR + RQ the shortest path from P to R to Q ?

A commonplace rendering of the problem is as follows :

AB is the bank of a river. A person wants to go from P to Q as fast as possible taking a bucketful of water from the river on the way. How would he make it ?

45. How is the problem solved ?

Let PP′ be drawn perpendicular to AB, P′ being as much behind AB as P is in front. Join P′ to Q. Then the line P′Q intersects AB in the required point R.

Here PR + RQ = P′R + RQ = P′Q

If R′ be any other point on AB, then PR + RQ will be smaller than PR′ + R′Q, because PR + RQ is equal to P′Q, and P′Q will always be less than the sum of the other two sides P′R′ and R′Q of the triangle P′R′Q.

Also, since $\angle 1 = \angle 3$, therefore, $90° - \angle 1 = 90° - \angle 3$, i.e., the angle of incidence = the angle of reflection, so that RQ must be the path of the reflected ray.

46. Raindrops and hailstones are spherical in shape ! Do they reveal any mathematical property ?

Yes, The sphere has the least surface for the volume it occupies.

47. What about the honeycomb, i.e., the wax structure of six-sided cells made by bees for honey and eggs ?

The cell of a bee is an upright prism upon a base which is a regular hexagon. The volume remains the same if the alternate corners B, D, F are cut off by planes through the lines AC, CE, EA meeting in a point V on the axis VN of the prism, and intesecting BB′, DD′, FF′.

The three planes make what is called a solid angle at V and for minimum surface make certain specific angles with the axis VN of the prism.

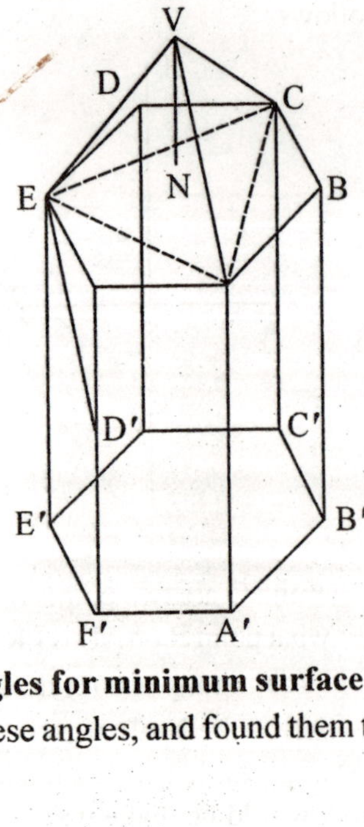

48. What is the measure of these angles for minimum surface ?

Maraldi was first who measured these angles, and found them to be 109°28′ and 70° 32′ respectively.

Re'aumur thought that this might be the form, which for the solid content, gives the minimum of surface. He, therefore, requested the geometer Koenig to examine the question mathematically.

As the result of Koenig's calculations agreed with Maraldi's measurement within 2', the geometer confirmed the conjecture as true.

49. The honeycomb is suppossed to be the product of the most unerring instinct of the bees, how then the difference between calculation and measurement, though not significant, can be accounted for ?

The answer lies in the following :

Maclaurin and L' Huillier also verified the conjecture by different methods, but showed that the difference of 2' was due to an error in the calculations of Koenig, and not to a mistake on the part of the bees !

50. What does it mean geometrically to say that the derivative does not exist ?

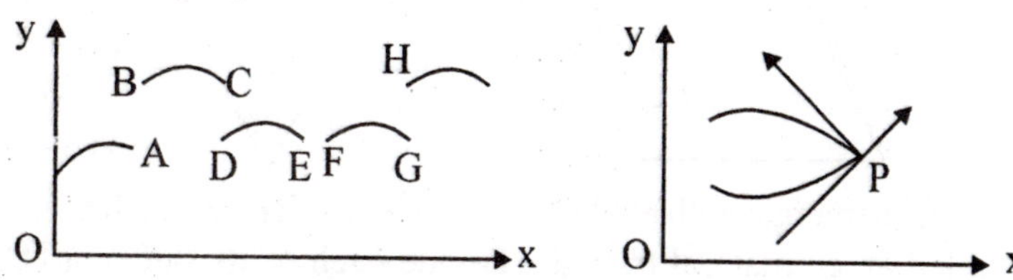

If a curve is discontinous at a point, i.e., there is a gap in the graph of the curve at some point, no tangent can be drawn to the curve at such a point, and the derivative there is said not to exist.

A, B, C, D, E, F, G, H are such points in the figure.

Again, if a curve is continuous but takes a shrap turn at some point. no single tangent can be drawn to the curve at that point,and the derivative there too is said not to exist.

P is such a point in the figure.

51. What is meant by saying that a continuous curve is ordinarily differentiable ?

It only means that a tangent can be drawn at every point of a continuous curve provided that the curve does not take a sharp turn at any point on it.

Since such points do not ordinarily occur on common continuous curves, the latter are said to be ordinarily differentiable.

But if such a a point is there on the curve, the curve is not differentiable at that point.

52. Could we have a continuous curve nowhere differentiable ?

Yes, if the curve had very frequent breaks i.e., sharp turns all along its length, it would be continuous but since a unique tangent could not be drawn to it at any point, such a curve would be continuous but nowhere differentiable.

53. How could such a curve be constructed ?

The following illustrates the method :

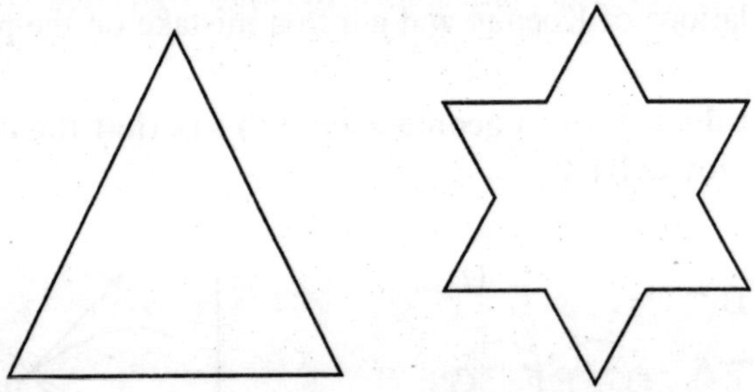

Take a triangle with all three sides of equal length. Such a triangle is called an equilateral triangle. Divide each of its sides into three equal parts and construct new equilateral triangles with peaks pointing out over the three central sections.

This gives a figure like a six pointed star. Each of the twelve sides of this star is further divided into three equal parts, again constructing equilateral triangles. After infinitely many divisions and construction of such triangles is obtained a curve at each point of which there will be a break or a prickle. The curve would, therefore, be nowhere differentiable.

54. What is meant by a continuous function ?

The general idea of a continuous function can be had from the fact that its graph is continuous, that is, its curve may be drawn without lifting the pencil from the paper.

The function $y = x^2$ is a continuous function. Its curve can be drawn without lifting the pencil from the paper.

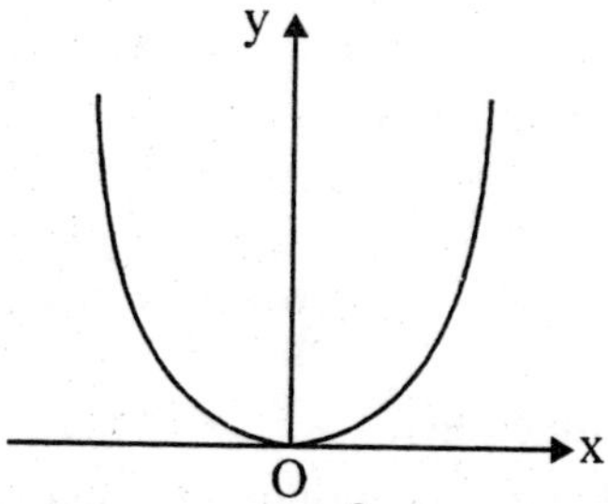

$y = x^2$

Similarly the following function is also continuous :

$y = x$, when $x \leq 1$,
$y = 1$, when $x \geq 1$.

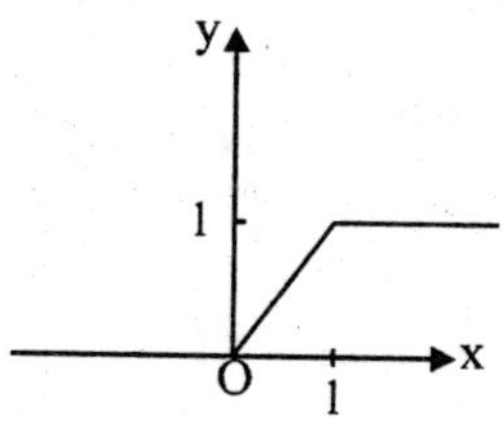

But the following is a discontinuous function.

$y = x$, when $x \leq 1$,
$y = 2$, when $x > 1$.

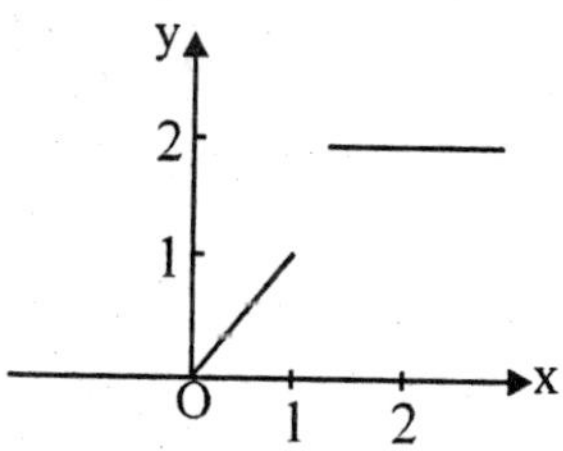

55. Is the function defined by $y = \frac{x^2 - 4}{x - 2}$ continuous ?

The function $y = \frac{x^2 - 4}{x - 2}$ is continuous everywhere except when $x = 2$, where it is not defined.

The function has a discontinuity at $x = 2$. But the same function redefined as :

$$\left.\begin{array}{ll} y = \dfrac{x^2 - 4}{x - 2}, & \text{when } x \neq 2, \\ y = 4, & \text{when } x = 2. \end{array}\right\}$$

is a continuous function.

It may be noted that the continuity is usually destroyed for those values of x for which the denominator vanishes.

56. How are plane curves studied in the calculus ?

In the calculus, plane curves are studied in terms of the following basic concepts :

Length, tangent and curvature.

57. How is the length of a curve defined ?

The length of a curve is the sum of the lengths of the broken lines inscribed in the curve under the condition that their vertices get closer and closer together on the curve.

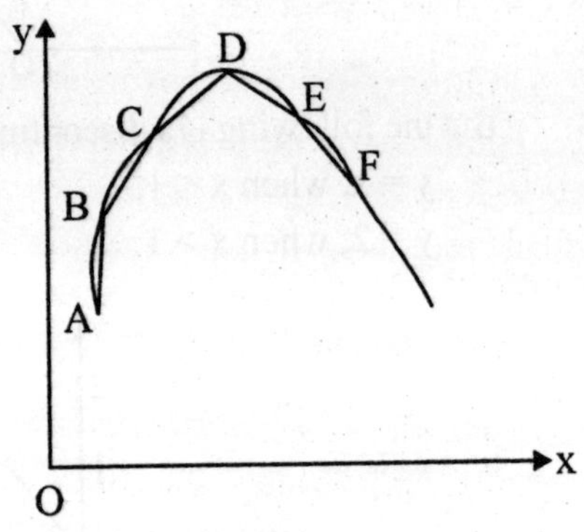

From the figure, length of the curve is the sum of the lengths AB, BC, CD, DE, EF, ..., as the vertices A, B, C, D, E, F, ... get closer and closer together on the curve.

58. What is meant by a closed curve of infinite length ?

Such a curve is constructible. The following illustrates the method.

Take a straight line of unit length. Divide it into two equal parts. Construct an isosceles triangle, which means, a triangle with two equal sides, on the left half, the altitude of the triangle being 1.

Next divide the right half $[\frac{1}{2}, 1]$ into two equal parts, and construct an isosceles triangle of altitude $\frac{1}{2}$ on the left part $[\frac{1}{2} \frac{3}{4}]$.

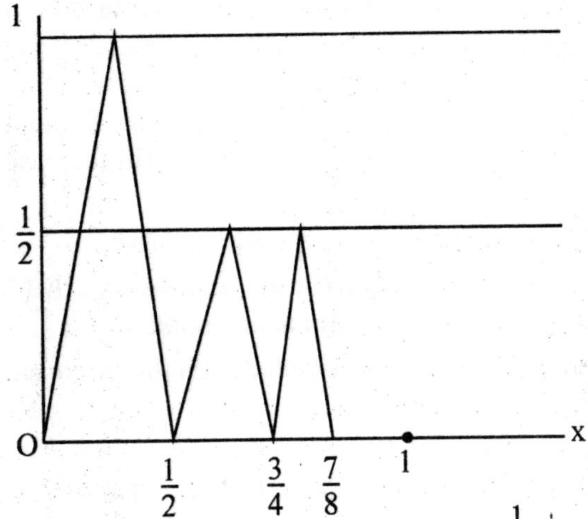

The next isosecles triangle, again with altitude $\frac{1}{2}$, is constructed on the part $\left[\frac{3}{4}, \frac{7}{8}\right]$.

The next four triangles are similarly constructed with altitude $\frac{1}{4}$, and the process continued indefinitely.

Since the length of the lateral side of the triangle is always greater than the altitude, the total length of the broken line cannot be less than the sum of the infinite series :

$$2 + \left(\frac{2}{2} + \frac{2}{2}\right) + \left(\frac{2}{4} + \frac{2}{4} + \frac{2}{4} + \frac{2}{4}\right) + \ldots\ldots$$

Since the sum of the numbers in each bracket is 2, and the number of brackets is infinite, the sum of the series is infinite and so is the length of our curve.

59. What is space-filling curve ?

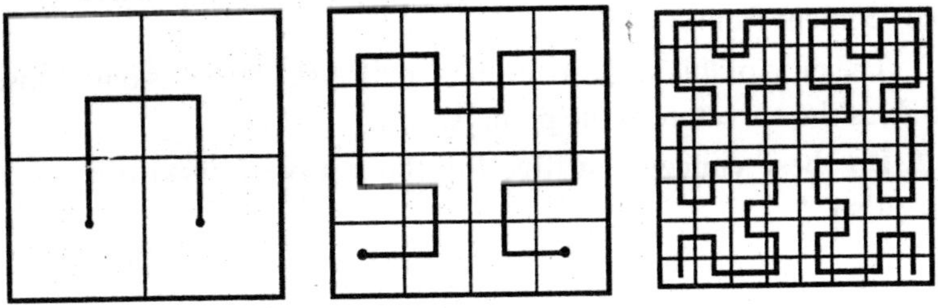

Take a square of side one unit length.

Subdivide this into four square regions as shown and connect in order the four midpoints of the regions. Each of these four square regions is subdivided into four parts, their centres connected as shown, always beginning in the lower left box and ending in the lower right box. If the process be continued indefinitely, the limiting curve will pass through every point in the square and fill the space within it completely.

The resulting curve is the space-filling curve.

This is also an example of a continues curve that is nowhere differentiable, i.e., has no tangent at any point.

The name of Hilbert is associated with this curve and is known as the Hilbert curve.

60. How is the tangent to a curve defined ?

If the chord AB takes on successively positions like AC, AD, AE, and converges to some limiting position AT, then the straight line in this position is called the tangent at the point A.

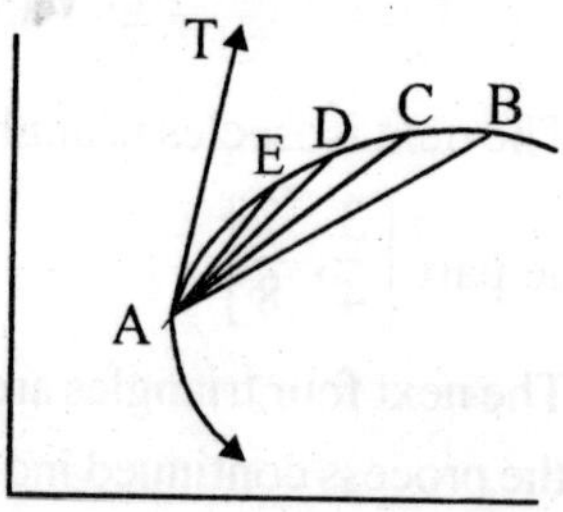

61. How is curvature defined ?

Curvature is the rate of turning of the tangent relative to the arc length.

Let A be a point on the curve and B a point near A. The angle between the tangents at the two points expresses how much the curve has changed direction in the segment from A to B.

The ratio of this angle to the arc length AB, when B is quite close to A, is the curvature at the point A.

62. Do these concepts suffice for the study of curves in space also ?

No. There are in addition, the osculating plane and the torsion.

63. What are they ?

The osculating plane of a space curve at a point is the limiting position of the plane through the tangent to the curve at the point.

Torsion may be taken as a measure of the rate at which the space curve is turning out of its tangent plane relative to the arc length.

64. What is meant by a function of two or more variables ?

So far we have dealt with functions of one variable only, but in practice it is often necessary to deal with functions depending on two, three or in general many variables.

For example, the area of a rectangle is a function,

$$A = xy,$$

where x is the length of its base and y its height.

The volume V of a rectangular parallelepiped is a function,

$$V = x\,y\,z,$$

where x, y, z are the length, breadth and the height of the parallelepiped.

65. What is a partial derivative ?

The extension of differentiation to functions of two or more variables leads to partial derivatives.

A function of x and y is differentiated partially with respect to x by considering y as temporarily constant.

If $u = f(x, y)$, the partial derivatives of u with respect to x and y are denoted by $\frac{\partial u}{\partial x}$ and $\frac{\partial u}{\partial y}$ respectively.

The symbol ∂ of partial derivatives is not a letter from any alphabet but a specially invented mathematical sign. It is pronounced 'delta'.

If $u = x^2 y + y^2 + 2x$, then

$$\frac{\partial u}{\partial x} = 2xy + 2, \text{ and } \frac{\partial u}{\partial y} = x^2 + 2y.$$

Such a derivative is interpreted as the rate of change of a function resulting from the change of one variable only.

66. What are higher partial derivatives ?

Higher partial derivatives for the above example are as follows :

$$\frac{\partial}{\partial x}\left(\frac{\partial u}{\partial x}\right) = \frac{\partial^2 u}{\partial x^2} = 2y$$

$$\frac{\partial}{\partial y}\left(\frac{\partial u}{\partial y}\right) = \frac{\partial^2 u}{\partial y^2} = 2,$$

$$\frac{\partial}{\partial x}\left(\frac{\partial u}{\partial y}\right) = \frac{\partial^2 u}{\partial x \partial y} = 2x$$

Functions of more than two variables are partially differentiated in a similar way.

Further, if $\frac{\partial^2 z}{\partial x^2}$, $\frac{\partial^2 z}{\partial x \partial y}$ and $\frac{\partial^2 z}{\partial y^2}$ be denoted by r, s and t respectively, we have :

For a maximum, $rt - s^2$ is positive, r and t both negative,

For a minimum, $rt - s^2$ is positive, r and t both positive.

If $rt - s^2$ is negative, we have a saddle point.

67. What is a saddle point ?

Consider two mountain peaks A and B on a range and two points C and D on different sides of the mountain range.

Strating from C, if a traveller arrives at E, he reaches at a maximum height in the direction in which he travels, but if he moves from the path either to left or right, he begins to ascend to higher ground.

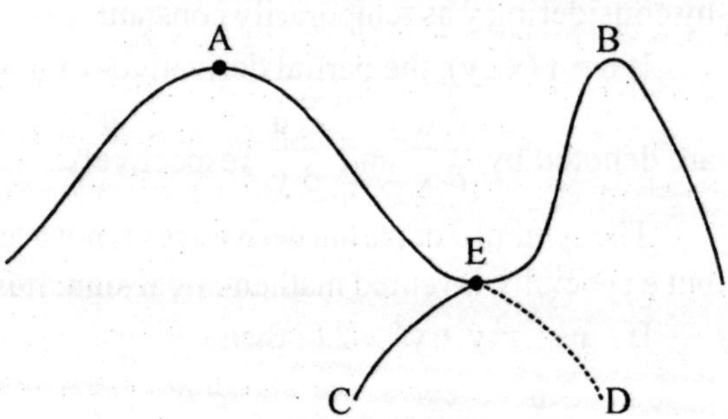

Thus the tangent plane to the surface at E is horizontal, but near the point the surface is partly above and partly below the tangent plane.

Such a point is called a saddle point due to its obvious similarity to the corresponding point on a saddle.

64. In what sort of problems is the integral calculus used ?

One use of the integral calculus is in determining the area enclosed by a curve.

The circle is the most familiar curve. Determining the area of a circle is easy. It can also be done by elementary geometry.

But the area of an irregular figure cannot be determined without the aid of the Integral Calculus.

Integration is also used to determine the length of a curve, the volume of a solid of revolution, the work done during the expansion of a gas, and in the solution of several such problems in science.

But the outstanding use of integration is summed up in the following:

When a process is under observation, generally nothing is known about it except the rate of the process. But a differential equation can be obtained easily, which on integration yields the knowledge of the whole process.

69. What is the Fundamental Theorem of the calculus ?

The Fundamental Theorem of the calculus states that differentiation and integration are inverse processes.

Mathematically, the theorem is stated as follows :

$$\int_a^b f(x)\ dx = [F(x)]_a^b = F(b) - F(a).$$

where a derivative of F(x) is f(x).

This extremely valuable discovery was one of the greatest among the contributions of Newton and Liebnitz.

70. In what way is the discovery valuable ?

Earlier, integrals were actually calculated by the summation method, which except for the simplest functions, is difficult and laborious.

Use of the inverse derivative for such calculations, therefore makes it an extremely valuable method. The method is much simple as well as general.

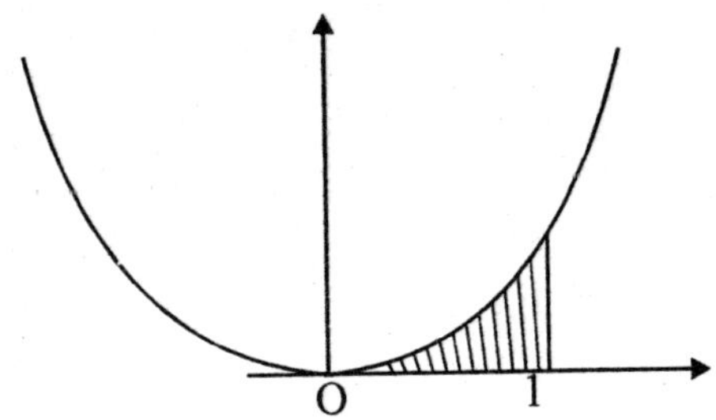

For example, if the equation to the curve be $y = x^2$ and the area under the curve and above the x-axis, terminated by vertical lines at $x = 0$ and $x = 1$ be required, the area is given by the integral

$$\int_0^1 x^2\,dx$$

This is best evaluated by observing that since the derivative of x^3 is $3x^2$, the integral of $3x^2$ is equal to x^3 and that of x^2 is equal to $\frac{x^3}{3}$.

$$\text{Thus, } \int_0^1 x^2\,dx = \left[\frac{x^3}{3}\right]_0^1$$

This is found by substituting 1 and 0 successively for x and subtracting the second value from the first, thus,

$$\left[\frac{x^3}{3}\right]_0^1 = \frac{1^3}{3} - \frac{0^3}{3} = \frac{1}{3} - 0 = \frac{1}{3}$$

Like Differential Calculus there are also standard results in the Integral Calculus. They are freely used for calculating integrals.

A few are being given here :

$$\int_a^b x^n\,dx = \left[\frac{x^{n+1}}{n+1}\right]_a^b = \frac{1}{n+1}\,(b^{n+1} - a^{n+1}),\ n \neq -1$$

$$\int_a^b \cos x\,dx = [\sin x]_a^b = \sin b - \sin a.$$

$$\int_a^b \sin x\,dx = [-\cos x]_a^b = -\cos b + \cos a.$$

$$\int_a^b \frac{1}{x}\,dx = [\log x]_a^b = \log b - \log a.$$

$$\int_a^b e^x\,dx = [e^x]_a^b = e^b - e^a.$$

These results are a direct consequence of the Fundamental Theorem of the calculus.

71. It is the summation method that enables one to calculate areas and volumes. How, then, calculation of the inverse derivative in stead does the trick ?

With integration viewed as summation, $\int_a^b f(x)\,dx$ represents the sum of the elements of an area under the curve y = f (x).

From the point of view of differentiation,

$$\int_a^b f(x)\,dx = [F(x)]_a^b = F(b) - F(a),$$

where a derivative of F(x) is f(x).

The two together show that the summation of the elements of an area under the curve y = f(x) is equivalent to the determination of a function F(x), whose derivative is f(x).

A similar argument holds good for the volume under a surface.

The summation method can, therefore, be replaced by the easier one of calculating the inverse derivative.

The underlying principle which makes this possible can be briefly stated as :

The ordinate of the curve is the derivative of the area.

Equivalently, the area is the integral of the ordinate.

y-coordinate of a point or the height of the curve at the point x is known as the ordinate.

72. Which functions are continuous ?

Some examples of continuous functions are x^n, sin x, cos x, a^x, log x, $\sin^{-1} x$, $\cos^{-1} x$ in the intervals for which they are defined.

In the case of a rational function, the continuity is usually destroyed for those values of x for which the denominator vanishes.

For example, the function $y = \frac{1}{x}$ is discontinuous at the point x = 0, where the denominator vanishes.

73. What is meant by saying that a continuous function is integrable ?

Since a continuous function can be represented by a continuous curve, the area enclosed by such a curve can be calculated, and the function is said to be integrable.

74. Are all continuous functions integrable ?

Yes, if a function is continuous in an interval, it is integrable in that interval.

The function $f(x) = \dfrac{1}{x-5}$ has a discontinuity at $x = 5$ so that it is integrable only in an interval which does not contain the point $x = 5$.

Thus the integral $\int_2^4 \dfrac{1}{x-5}\,dx$ can be evaluated but not the integral $\int_4^6 \dfrac{1}{x-5}\,dx$.

75. To be integrable, is it necessary for a function to be continuous over the interval ?

No. Even when the function is not continuous in an interval, but the interval can be divided into a finite number of sub-intervals, in each of which the function is continuous except at the points of juncture, the function is still integrable.

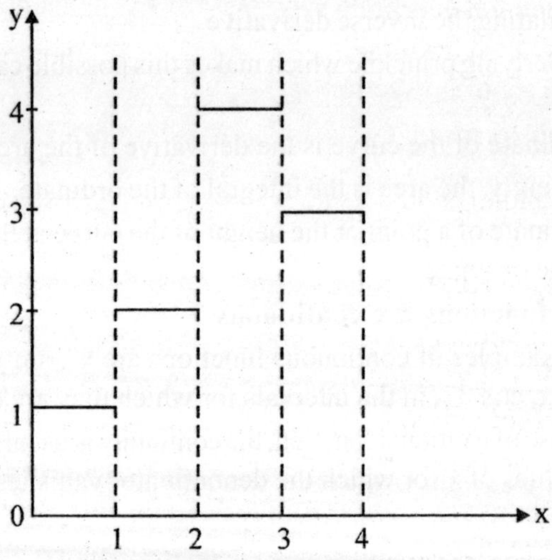

Let a function be defined as :

$$y = 1, \text{ when } x \le 1,$$
$$y = 2, \text{ when } 1 < x \le 2,$$
$$y = 4, \text{ when } 2 < x \le 3,$$
$$y = 3, \text{ when } 3 < x \le 4.$$

Here x = 1, x = 2, and x = 3 are the points of discontinuity.

The function is not continuous in the whole interval (0,4), but it is piece-wise continuous in the sub-intervals (0, 1) (1, 2), (2, 3) and (3,4).

It is, therefore, integrable in the sub-intervals separately, and thus supposed to be integrable in the interval (0, 4).

$$\text{Actually, } \int_0^4 y\ dx = \int_0^1 y\ dx + \int_1^2 y\ dx + \int_2^3 y\ dx + \int_3^4 y\ dx,$$

$$= \int_0^1 1\ dx + \int_0^2 2\ dx + \int_2^3 4\ dx + \int_3^4 3\ dx,$$

$$= 1 + 2 + 4 + 3,$$

$$= 10.$$

76. What is Riemann Integral ?

The ordinary integral $\int_a^b f(x)\ dx$ is known as the Riemann integral in honour of the great German mathematician Riemann who developed the concept.

77. When does a function fail to be Riemann-Integrable ?

This many happen when the function becomes unbounded.

For example, the function $\frac{1}{x}$ becomes unbounded near x = 0, so that $\frac{1}{x}$ is not integrable in any interval containing the origin.

A function may also fail to be Riemann-integrable if it has an infinite number of discontinuties.

For this consider the function defined over the interval [0, 1] as

$f(x) = 0$, where x is rational,

$f(x) = 1$, when x is irrational.

Between any two rational numbers, no matter how close they are, there is an infinite number of irrational numbers,and vice versa, so that the function continually fluctuates in value between 0 and 1 over the entire interval. The function is therefore totally discontinuous, the discontinuities being infinite in number.

The function therefore has no integral in the Riemann sense.

78. Does it mean that the function is integrable in any other sense ?

Yes, the function has a Lebesgue integral, which is a generalisation of the Riemann integral.

Lebesgue integral can be applied to totally discontinuous functions, whereas to be Riemann integrable, the function must somehow satisfy the continuity property even if it be in the sub-intervals separately.

79. Integrating a totally discontinuous function appears to be apparently an impossible task. How does Lebesgue integral accomplish it ?

In the Riemann sense, integration is always over intervals, while in the Lebesgue sense it is over sets.

It is the replacement of intervals by sets that makes it a very powerful tool and greatly enhances the scope of integration because it can be applied to a large class of such functions as are defined only over sets.

80. But whereas the length of an interval is an obvious entity, what could be the length-analogue of a set ?

The length-analogue of a set is called its measure.

Consider again the function defined over the interval [0, 1] as

$f(x) = 0$, when x is rational,

$f(x) = 1$, when x is irrational.

The interval [0, 1] is composed of rational and irrational points.

The interval may be looked upon as the union of two sets of points, the set of rational points and the set of irrational points sothat the measures or the length-analogues of the two sets are together equal to the length of the interval.

It can be shown that in any interval the irrational points are far more numerous than the rational points, and occupy the entire length of the interval.

The length-analogue or the measure of the set of irrational points in the interval is thus equal to the length of the interval, i.e., equal to 1.

The measure of the set of rational points in the inteval is, therefore, zero.

81. How does the function under reference become integrable in the Lebesgue sense ?

At the rational points the value of the function is zero so that the contribution of the set of rational points to the integral is zero, and the function need be integrated only over the set of irrational points in the interval [0, 1].

At the irrational points the value of the function at each such point is 1, and the length-analogue of the set of irrational points being 1, the value of the integral over the set of irrational points is 1 x 1, which is equal to 1.

The function, therefore has a lebesgue integral over the interval [0,1], its value being 1.

82. What is Fourier's Theorem or Fourier series ?

Fourier showed that ordinarily *any* arbitrary function can be expressed as a sum of sines and cosines in the following form :

$$f(x) = a_0 + a_1 \cos x + a_2 \cos 2x + a_3 \cos 3x + ... + a_n \cos nx + ...$$
$$+ b_1 \sin x + b_2 \sin 2x + b_3 \sin 3x + ... + b_n \sin nx + ...$$

This is known as the Fourier series.

The constants $a_0, a_1, a_2, a_3, ... a_n ; b_1, b_2, b_3, ... b_n$ can be determined by the following formulae :

$$a_0 = \frac{1}{2\pi} \int_{-\pi}^{\pi} f(x)\, dx,$$

$$a_n = \frac{1}{\pi} \int_{-\pi}^{\pi} f(x) \cos nx\, dx, n = 1, 2, 3, \text{ etc.}$$

$$b_n = \frac{1}{\pi} \int_{-\pi}^{\pi} f(x) \sin nx\, dx, n = 1, 2, 3, \text{ etc.}$$

These are called Fourier Integrals.

This remarkable discovery has been hailed as one of the most astonishing results in the whole range of human knowledge.

83. Wherein lies the secret of this miraculous result ?

The secret lies in the very nature of thc simple periodic functions.

A function is said to be periodic if its graph is a repeating pattern. Sine and cosine of trigonometry are two such simple functions. Since the graphs of sin x and cos x repeat after every 360° or 2π radians or simply 2π, sin x and cos x are periodic functions, their period being 360° or 2π.

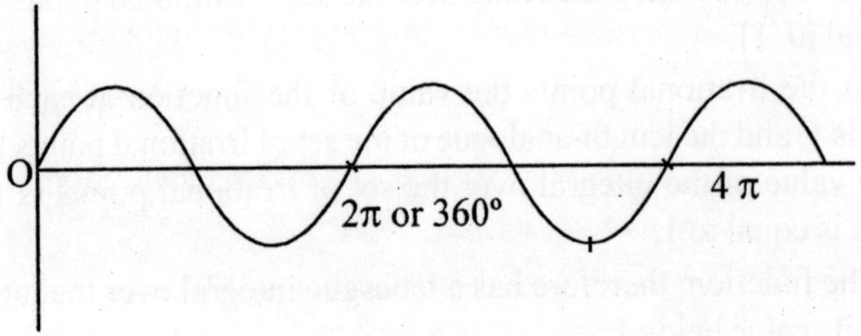

Sine curve

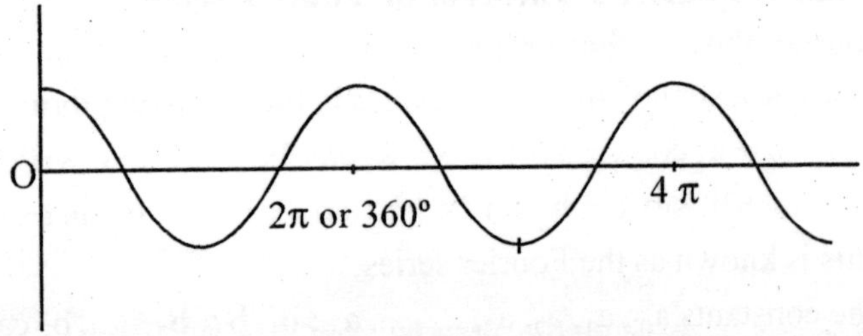

Cosine curve

As we go along the sequence, the periods of sin 2x, sin 3x, sin 4x, ... and also of cos 2 x, cos 3 x, cos 4 x, ... get shorter and shorter.

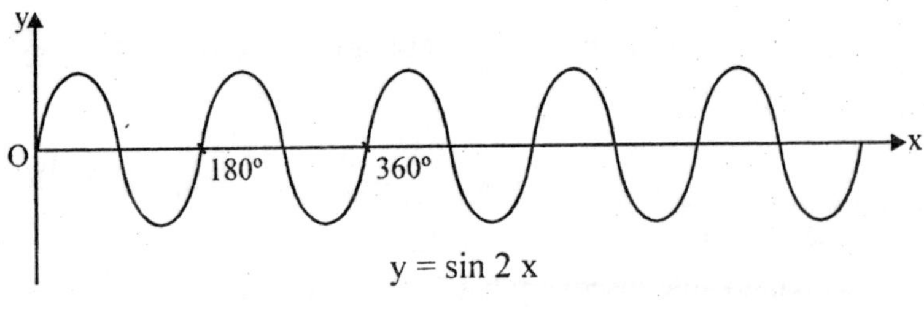

y = sin 2 x

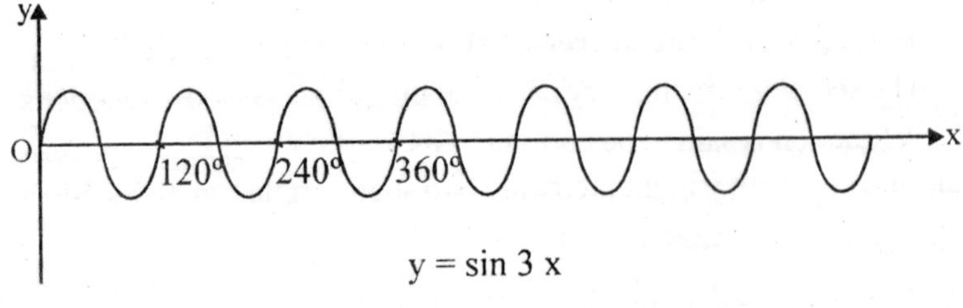

y = sin 3 x

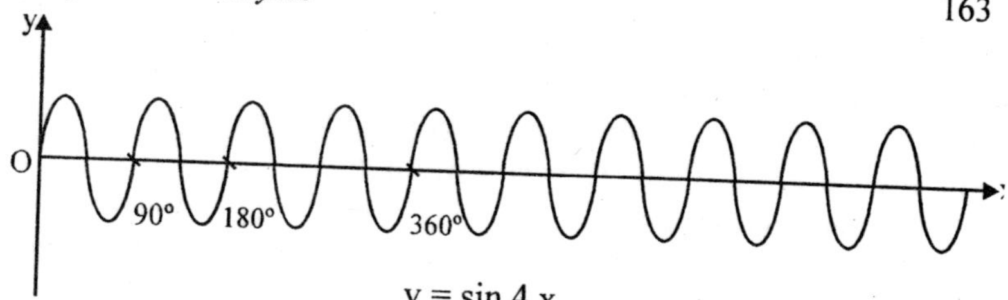

y = sin 4 x

Thus,

the period of sin 2 x is half of 360° i.e. 180°,

the period of sin 3 x is one-third of 360°, i.e. 120°,

the period of sin 4 x is one-fourth of 360°, i.e. 90°,

and so on.

These are also respectively the periods of cos 2x, cos 3x , cos 4x etc.

Hence as the angle goes on increasing, its period goes on decreasing so that any periodic graph, howsoever complicated, can be made up by adding a sufficient number of the graphs of the form a_1 sin x, a_2 sin 2x, a_3 sin 3x, ... and b_1 cos x, b_2 cos 2x, b_3 cos 3x,....

Consequently almost any function can be expressed in the form of a Fourier series.

84. What is remarkable about such a series representation of a function?

The remarkable thing is that the series represents the function even when the function has a finite number of discontinuities or a finite number of points at which the derivative does not exist.

It can also be used to represent functions that ordinarily are represented by different expressions in different parts of the interval.

These turn out to be extremely valuable achievements of the fourier series for the following reason :

From the viewpoint of mathematicians, study of a function turns out to be fruitful only if it is well-behaved, which means that it should be continuous and differentiable almost everywhere.

But with the possibility of such a series representation, functions with considerably greater generality can be studied and are no longer required to be of the well-behaved form. They can even be reasonably discontinuous and non-differentiable.

85. What is the advantage of the Fourier series representation of a function ?

One great advantage of expressing a function in terms of sines and cosines lies in the simple behaviour of these functions under the various operations of analysis, especially differentiation.

86. Where is the Fourier Series applied ?

It finds application in all repeating or periodic phenomenon.

Many natural and artificial phenomenon occur in cycles that repeat constantly.

Natural phenomena like the seasons, the Sunrise and the Sunset, Sunspots, planetary orbits, the tides and several others are periodic or approximately so.

Some other examples of periodic phenomena are light waves, sound waves, elctromagnetic waves, alternating currents, artificial satellites, vibrating strings and business cycles.

Fourier analysis is indispensable where wave motion underlies the pattern of events.

87. Why is there so much emphiasis on periodic phenomena ?

Because periodic phenomena are very common in our daily lives, although we seldom notice them. In nearly equally spaced intervals of time we repeat the same things over and over again in nearly the same way.

Two phenomena closest to us - the very act of breathing, and the heartbeats, are also periodic.

88. Can the function f (x) = x in the interval – π ∠ x ∠ π be represented as a sum of the sines of angles ? If yes, what is such representation ?

Yes, the given function can be expressed as a Fourier series. It is expressible as a sum of the sines of angles. Thus,

$$x = 2 \left[\sin x - \frac{1}{2} \sin 2x + \frac{1}{3} \sin 3x - \frac{1}{4} \sin 4x + ... \right].$$

89. What is the role of continuity in the physical sciences ?

It is well known that the so called dense media, namely solids, liquids and gases ; in particular, metals, water, and air are in fact accumulation of a large number of separate particles in motion.

These particles and the distances between them are so small in comparison with the dimensions of the media in which the phenomena take place that the medium may be considered as continuously distributed over the occupied space.

Many physical sciences, for example, hydrodynamics, aerodynamics, and the theory of elasticity are based on such assumption.

The mathematical concept of continuity, therefore, plays a very important role in such sciences and in many others.

90. What are multiple integrals ?

Multiple integrals are repeated integrals, and are used when the function to be integrated consists of two or more independent variables.

They are evaluated by successive single integrations.

The following is an example of a multiple integral of order two, also called a double integral :

$$\int_0^1 \int_0^{x^2} xy \, dx \, dy$$

91. How is it evaluted ?

It is done as follows :

$$\text{Let } I = \int_0^1 \int_0^{x^2} xy \, dx \, dy$$

$$= \int_0^1 \left[\int_0^{x^2} xy \, dy \right] dx \qquad .....(1)$$

First the integral within the brackets is evaluated by treating x as constant.

$$\text{Let } I_1 = \int_0^{x^2} xy \, dy$$

$= x \int_0^{x^2} y \, dy$, x is taken out of the integral sign as it is being treated as constant.

$$= x \left[\frac{y^2}{2} \right]_0^{x^2}, \quad \because \text{ the integral of y is equal to } \frac{y^2}{2}$$

$$= x \left[\frac{(x^2)^2}{2} - 0 \right] = \frac{x^5}{2}$$

Thus, $I = \int_0^1 \frac{x^5}{2}\, dx$, by (1)

$$= \frac{1}{2}\left[\frac{x^6}{6}\right]_0^1 = \frac{1}{2}\left[\frac{1^6}{6} - 0\right] = \frac{1}{12}$$

92. What are the main branches of analysis ?

The following are the main branches of analysis :

Differential and Integral calculus,
Theory of Infinite series,
Theory of Differential Equations,
Differential Geometry,
Calculus of variations,
Theory of Functions of a Real Variable,
Theory of Functions of a Complex Variable,
Approximation of Functions,
Functional Analysis,
Theory of Distributions.

93. What is a differential equation ?

If a relation in the form of an equation, contains a derivative of a function, it is a differential equation. it may contain an independent variable x, a dependent variable y, and the derivative $\frac{dy}{dx}$.

Thus, $(1 - x)\frac{dy}{dx} = 1 + y$ is a differential equation.

94. How is a differential equation formed ?

A differential equation is formed by the elimination of arbitrary constants in a relation. Thus the relation

$$y = A \sin x + b \cos x \quad ....(1)$$

has two arbitrary constants, A and B, and to eliminate them, three equations are required. Of these three equations, one is given, namely (1), and the two others needed are obtained by successive differentiation of (1). Thus,

$$y = A \sin x + B \cos x,$$

$$\frac{dy}{dx} = A \cos x - B \sin x\,,$$

$$\frac{d^2y}{dx^2} = -A\sin x - B\cos x,$$
$$= -(A\sin x + B\cos x),$$
$$= -y, \because A\sin x + B\cos x = y, \text{ by (1)},$$

so that $\frac{d^2y}{dx^2} + y = 0$ is the required differential equation.

95. What is an ordinary differential equation ?

Ordinary differential equations are those in which all the differential coefficients are with respect to a single independent variable. Thus

$$x\frac{d^2y}{dx^2} + 2x\frac{dy}{dx} + 2y = 0$$

is an ordinary differential equation because it involves a single independent variable x.

96. What is a partial differential equation ?

Partial differential equations are those which contain one or more partial derivatives, and are therefore concerned with at least two independent variables. Thus,

$$xz\frac{\partial z}{\partial x} + yz\frac{\partial z}{\partial y} = xy$$

is a partial differential equation.

If x and y be taken as independent variables and z the dependent variable, the partial differential coefficients $\frac{\partial z}{\partial x}$ and $\frac{\partial z}{\partial y}$ are denoted by the symbols p and q respectively.

In partial differential equations of second and higher orders, the partial derivatives $\frac{\partial^2 z}{\partial x^2}$, $\frac{\partial^2 z}{\partial x\,\partial y}$ and $\frac{\partial^2 z}{\partial y^2}$ are denoted by r, s and t respectively.

97. What is meant by solving a differential equation ?

To solve a differential equation means to find y as a function of x.

In ordinary algebra solving an equation means determining the unknown quantity x.

In the case of differential equation, the function f(x), i.e., y is the unknown entity.

Formal methods of solving various types of differential equations have been developed and mostly completed by the middle of the eighteenth century.

98. Can all differential equations be solved ?

No. Not all differential equations can be solved by the classical methods.

When classical devices fail, final resort may be made to numerical, graphical or mechanical methods. High speed electronic computers have made it possible to complete the solution of many differential equations, especially non-linear ones, which would otherwise be extremely difficult to solve.

99. What is meant by the theory of differential equations ?

The theory of differential equations is concerned with the existence and validity of solutions rather than with the actual task of solving the equations. This aspect of the study is of main concern to the mathematicians dealing with the foundations of mathematics.

However, applied mathematicians and scientists are more interested in obtaining a solution.

100. Where are differential equations mostly used ?

They are mostly used in applied mathematics. Dynamical situations nearly always involve rates, such as velocities and accelerations, and these are expressed as derivatives, hence the use of differential equations.

Familiar examples of dynamical situations are also found in flows of air, water and electricity.

A great many physical, chemical and economic processes also fit into this frame.

101. What is Laplace's equation ?

The partial differential equation

$$\frac{\partial^2 V}{\partial x^2} + \frac{\partial^2 V}{\partial y^2} + \frac{\partial^2 V}{\partial z^2} = 0$$

is known as Laplaces' differential equation in three dimensions.

102. Where is this equation used ?

This very famous equation is most frequently met with in investigations in applied mathematics, especially where the theory of potential is involved.

For example, if V be the Newtonian potential due to an atracting mass, at any point P (x, y, z) not forming a part of the mass itself, V satisfies Laplace's equation.

Again, if V be the electric potential at any point (x, y, z), where the electric density is zero, V satisfies Laplace's equation.

Likewise, if a body be in a state of equilibrium as to temperature, V being the temperature at any point, and $\frac{dV}{dt} = 0$, then V satisfies the Laplace's equation.

If f (x, y, z) denote any value of V that satisfies the Laplace's equation, then f (x, y, z) = c in the first two examples is called as equipotential surace, and in the third an isothermal surface.

Still another example : In the case of fluid flow, if V be the velocity potential, then V satisfies the Laplace's equation.

Also, in the investigation of the filtration of a liquid through a porous medium, we arrive at the Laplace's equation.

The equation also occurs in the theories of light, sound and elasticity.

103. Wireless waves were first predicted by mathematical equations. What are those equations ?

Mathematical prediction of wireless waves is embodied in Maxwell's equations. They are also partial differential equations.

The following notation is used in the equations.

K is the specific inductive capacity, μ the permeability, c the velocity of light in free space ; E_x, E_y, E_z, are the components of the electric intensity E in directions parallel to the axes of coordinates at the point (x, y, z); H_x, H_y, H_z are the components of the magnetic intensity H.

The equations for a homogeneous isotropic medium containing no free charge are

$$\frac{\partial E_x}{\partial x} + \frac{\partial E_y}{\partial y} + \frac{\partial E_z}{\partial y} = 0,$$

$$\frac{\partial H_x}{\partial x} + \frac{\partial H_y}{\partial y} + \frac{\partial H_z}{\partial y} = 0 ;$$

$$\frac{\partial H_z}{\partial y} - \frac{\partial H_y}{\partial z} = \frac{k}{c} E_x ,$$

$$\frac{\partial H_x}{\partial z} - \frac{\partial H_z}{\partial x} = \frac{k}{c} E_y,$$

$$\frac{\partial H_y}{\partial x} - \frac{\partial H_x}{\partial y} = \frac{k}{c} E_z;$$

$$\frac{\partial E_z}{\partial y} - \frac{\partial E_y}{\partial z} = -\frac{\mu}{c} H_x,$$

$$\frac{\partial E_x}{\partial z} - \frac{\partial E_z}{\partial x} = -\frac{\mu}{c} H_y,$$

$$\frac{\partial E_y}{\partial x} - \frac{\partial E_x}{\partial y} = -\frac{\mu}{c} H_z.$$

By eliminating either H or E from these equations the wave equation of mathematical physics results. The wave equations for E and H show that electromagnetic disturbances are propagated in vacuum as waves travelling with the velocity of light. This was the basis for the prediction of wireless waves.

104. When was this prediction made ?

Maxwell's mathematical prediction of wireless waves was made in 1864.

Experimental verification of the waves, twenty four years later, by Hertz came in 1888.

Marconi made the commercial beginning by sending wireless signals across the English Channel in 1899. Since then have sprung the whole wireless and radio and television industries of today.

105. What is wave equation ?

The partial differential equation :

$$\frac{\partial^2 \Psi}{\partial x^2} + \frac{\partial^2 \Psi}{\partial y^2} + \frac{\partial^2 \Psi}{\partial z^2} = \frac{1}{c^2} \frac{\partial^2 \psi}{\partial t^2}$$

is known as the wave equation.

In the theory of sound, this eqution is satisfied by the velocity potential in a perfect gas.

In the theory of elastic vibrations, it is satisfied by each component of the displacement.

In the theory of electric or electromagnetic waves, it is satisfied by each component of the electric or magnetic (force) vector.

The constant c is the velocity of propagation of the periodic distrurbance.

106. What is an integral equation ?

Integral equations involve integrals rather than derivatives. Thus and integral equation is an equation in which the unknown function occurs under an integral sign.

For example, the equation

$$f(x) = \int_{-\infty}^{\infty} \cos(xt)\,\phi(t)\,dt,$$

where f is an even function, is an integral equation.

Under certain conditions, a solution is

$$\phi(x) = \frac{2}{\pi}\int_{0}^{\infty} \cos(u\,x)\,f(u)\,du.$$

Another integral equation is the following :

$$f(x) = \int_{a}^{b} K(x, t)\,y(t)\,dt,$$

where f and K are two given functions and y is the unknown function.

The subject of integral equations is considered less vast than that of differential equations and not as important.

107. What is meant by the function of a complex variable ?

W = f (z) is called a function of the complex variable z.

The letter z is employed to denote x + iy, and is called a complex variable, x and y being real numbers, and i stands for $\sqrt{-1}$.

108. What does complex variable analysis deal with ?

It deals with the calculus of the functions of a complex variable.

109. Which type of functions are dealt with in this study ?

This study is concerned mainly with functions that have a derivative at every point of a given domain of values for z. Such functions are called analytic or regular functions.

110. In what way does the function of a complex variable differ from the function of a real variable ?

If a function of a complex variable has a first derivative, it has derivatives of all orders. But this is not necessarily true for a function of a real variable.

This makes up the most characteristic difference between the two types of functions.

111. What does Cauchy's Integral Formula imply ?

Cauchy showed that if a function of a complex variable has a derivative, then its value at a given point can be expressed as an integral of its values at other points.

Symbolically the formula can be stated as follows :

If f (z) is analytic within and on a closed contour C; and a is any point within C, then

$$f(a) = \frac{1}{2\pi i}\int_C \frac{f(z)}{z-a}\,dz,$$

where C stands for any closed path of integration enclosing the point z = a.

112. What is meant by the uniqueness property of an analytic function?

The property is expressed in the following theorem :

It two analytic functions agree on some curve C in a domain D, they agree on the entire domain.

This is one of the most remarkable properties of analytic functions.

113. What happens in the case of a real variable ?

Consider the case of a real variable, and let $f(x) = e^{-1/x}$ for $x > 0$.

The function is infinitely differentiable, but may be defined to be zero or anything along negative x-axis.

Thus, no unique function is obtained.

114. How is an analytic function related to fluid-flow ?

If we choose an arbitrary differentiable function of a complex variable, its real and imaginary parts may be taken as the velocity potential and the stream function of the flow of an incompressible fluid.

For example, if $f(z) = z^2$ be an analytic function, we have

$$\begin{aligned} z^2 &= (x + i\,y)^2 \\ &= x^2 - y^2 + i\,2\,xy \\ &= u + i\,v \text{ (suppose).} \end{aligned}$$

Thus $u = x^2 - y^2$, and $v = 2\,xy$ are respectively the real and imaginary parts of f (z).

They may be taken respectively as the velocity potential and the stream function of a fluid flow.

The real and the imaginary parts, each of them, also satisfy the Laplace's equation in two dimensions, viz.,

$$\frac{\partial^2 V}{\partial x^2} + \frac{\partial^2 V}{\partial y^2} = 0$$

115. How is the fluid flow around bodies of arbitrary shape determined ?

With the methods of the functions of a complex variable, flow around a body of arbitrary shape can be reduced to the case of the flow around a cylinder. Since the nature of flow around a cylinder is completely known, flow around a body of any shape can be determined.

116. What is the discovery of Zukovskii ?

The remarkable discovery of Zukovskii consists of the fact that the existence of circulation in the flow causes a lifting force on the wing of an aeroplane in a direction perpendicular to the velocity of the oncoming flow and is equal in magnitude to the quantity.

$$\rho\, a\, \Gamma ,$$

where ρ is the density of the medium, a the velocity of the oncoming flow, and Γ the circulation.

This theorem about the lifting force on a wing is basic for all contemporary aerodynamics.

117. What is Riemann's zeta function ?

Riemann's zeta function, denoted by the letter ζ of the Greek alphabet, is defined by the series

$$\zeta = f(z) = \sum_{n=1}^{\infty} n^{-z}$$

$$= \frac{1}{1^z} + \frac{1}{2^z} + \frac{1}{3^z} + \frac{1}{4^z} + \ldots\ldots ,$$

where $z = x + i\,y$, x and y are real and $i = \sqrt{-1}$.

118. What is meant by the zeros of the zeta function ?

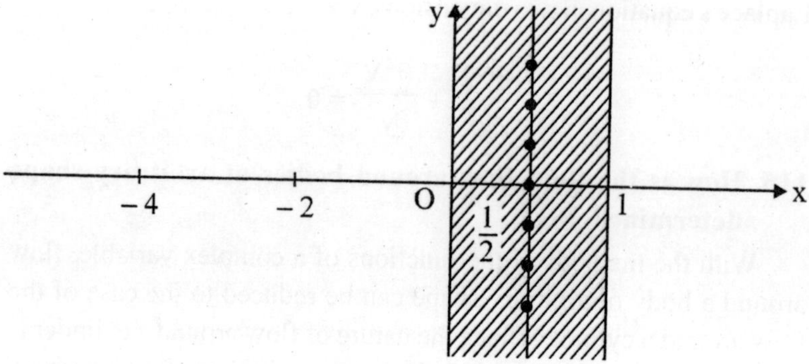

● = Conjectured solutions to Reimann's equation.

Those numbers for which zeta function vanishes are called zeros of the zeta function.

They are the solutions of the equation

$$1+\frac{1}{2^z}+\frac{1}{3^z}+\frac{1}{4^z}+......=0.$$

The zeta function has no zeros for $x > 1$ and is analytic there.

Along negative part of the real axis, the zeta function has zeros at – 2, – 4, ...

It is known that all other zeros of zeta function lie somewhere inside the shaded "critical strip" defined by $0 \leq x \leq 1$.

119. What is Riemann's hypothesis about the zeros of the zeta function?

Riemann made a conjecture that zeros of the zeta function inside the critical strip all lie on the central line of this strip, i.e., on the line $x = \frac{1}{2}$.

It is highly probable that Rimann's conjecture is correct but since it has not yet been proved, the conjecture is called the Riemann hypothesis.

That an infinite number of zeros lie on this line was proved by G. H. Hardy.

It has also been proved that at least one third of all the zeros lie on the critical line.

Using a computer it has also been shown that the first three million zeros of the zeta function lie on the central line of the strip.

Three million cases without a single exception might seem strong evidence but such finite reasoning is quite unreliable in the domain of the infinite.

120. If the conjecture is proved correct, what consequence will it lead to ?

A proof of the conjecture would have far-reaching consequences for the theory of prime numbers since the proof would actually imply sharp improvements in many of the known results about prime numbers.

121. What is the calculus of variations about ?

It is the study of the theory of maxima and minima of definite integrals, the problem being to determine the dependent variable so that the integral will be a maximum or a minimum.

A simple problem is to determine y so that the integral

$$\int_a^b f(x, y, \frac{dy}{dx})\,dx$$

is either a maximum or a mimimum.

122. But problems regarding maximum and minimum are studied in the differential calculus also !

In the calculus of variations the problems are of more general character than those in the differential calculus, where they require the location of a point with specified properties, while in the calculus of variations, a curve or surface is sought.

For instance, in the differential calculus, the problem of minimum in its simplest form requires us to minimize a function of one variable. More difficult problems regarding maximising or minimising of functions of more variables also involve only a finite number of the variables.

123. What sort of problems are dealt with in the calculus of variations ?

Consider the problem of determining the curve down which a ball will roll from a fixed point to a lower one not directly below it in the shortest time. The time required depends not upon one point or a few points of the curve, but upon all the points of the curve, which are infinite in number.

124. Which curve meets this requirement ?

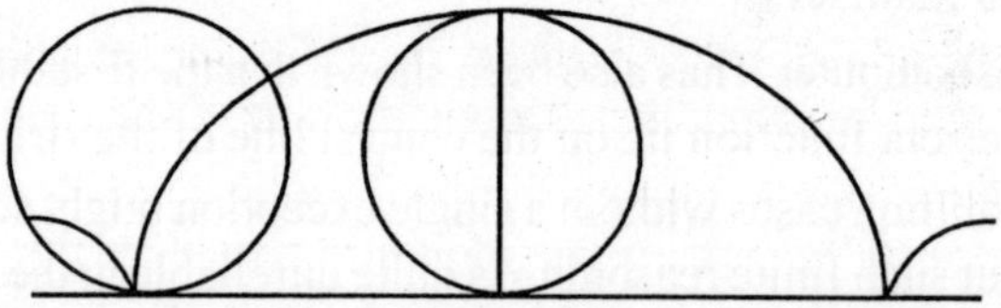

Cycloid

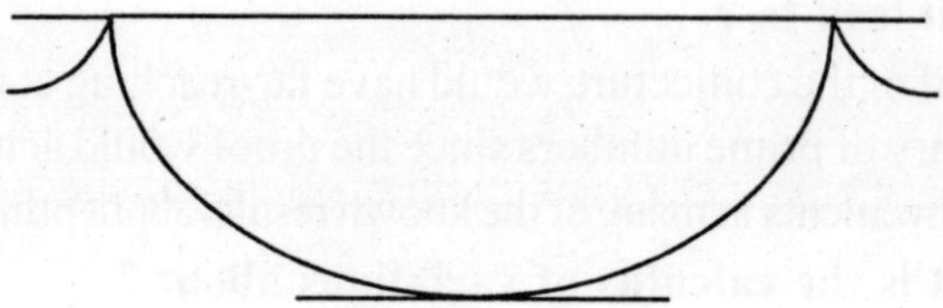

Inverted cycloid

It is the curve called inverted cycloid.

The cycloid is traced out by a point on a circle that rolls along a straight line.

The experiment can, of course, be performed with an inverted cycloid because the ball has to fall down the curve.

125. Why is the cycloid called "the Helen of geometers" ?

During the 17th century mathematicians were engaged in studying the cycloid and its properties, viz., area under one arch of the curve, area of the generating circle, volume generated by the revolution of the area about the axis of symmetry or about the tangent at the vertex, and such other matters.

Priorities regarding discovery of these and other properties turned out to be the matter of frequent quarrels among mathematicians so that the phrase "the Helen of geometers" came to be applied to cycloid.

126. What are the applications of the calculus of variations ?

The subject has applications in Economics, Business and other practical affairs because in these areas maximum profit and minimum cost and effort are to be secured. The subject has, therefore, great practical use, and is also of theoretical interest.

127. What is the Theory of Distributions ?

It is a generalization of the concept of differentiation. In differential calculus, especially in its applications, there is a constant temptation to differentiate functions that do not have derivatives. One way to avoid doing this is to take the precautionary measure of only working with differentiable functions, but there are other ways also.

One very effective artifice has been to extend differential calculus from differentiable functions to a kind of generalized functions called distributions in such a way that essentially all the algebraic rules still hold good.

128. What prompted such a generalization ?

The necessity arose because of problems in probability.

Classical analysis had been concerned with continuous functions, whereas problems in probability generally involve discrete cases.

A closer association of analysis and probability, therefore, resulted on account of the concept of measure theory, the extension of the concept of integration, and the generalization of the concept of differentiation through the theory of distributions.

129. Who introduced this generalization ?

The theory of distributions was worked out by Laurent Schwartz of the University of Paris at about 1950.

130. What is achieved by this generalization ?

Schwartz developed an appropriate definition of the derivative of a distribution such that the derivative of a distribution always is itself a distribution. This provides a powerful generalization of the calculus with immediate applications to probability theory and physics.

Its starting point is the formula for integration by parts.

131. What are Generalized Functions ?

Distributions in this sense are also known as Generalized Functions.

132. What is Real Analysis ?

It is a subdivision of analysis that is dependent upon or suggested by the properties of the real number system.

133. What topics are studied in real analysis ?

Some topics of main interest are :

Continuity, differentiability and integrability of functions of one and more than one variable. Convergence of infinite series is another main topic.

134. What were the early worries of analysis ?

By about 1800, several concepts making up analysis were in bad shape.

The concept of function was not clear.

Indiscriminate use of infinite series had produced paradoxes.

Fourier's representation of functions in terms of sines and cosines posed serious difficulties and disagreements.

And the notions of the derivative and the integral were not properly defined.

135. How was this state of affairs set right ?

The following significant developments helped to clear the picture.

Calculus and its extensions were freed from all dependence upon geometrical notions. Dependence upon motion and such other views were also done away with.

The concept of continuity was put on a sound basis.

The discovery that continuous functions need not have derivatives came as a surprise and a shock, but led to greater clarity of thought.

Surprising also was the discovery that discontinuous functions can be integrated.

With refinements provided by the works of Riemann and Lebesgue, new light was shed on which functions were integrable.

136. How did Real Analysis come into being ?

Formerly, only those functions were studied which satisfied the requirement of differentiability. This restricted the choice in the class of functions.

The study of functions was continued in the twentieth century and resulted in the development of a new branch of mathematics known as the theory of functions of a real variable, also known as Real Analysis.

137. How is Functional Analysis different from classical analysis ?

In classical analysis a variable is treated as a magnitude but in Functional Analysis the function itself is regarded as a variable.

What is under study is not a particular function, but a whole collection of functions having the same common property.

138. What could be such functions ?

The following are a few examples :

1. The collection of all continuous functions.
2. The collection of all curves on a surface, thereby defining the properties of the separate curves in their relation to other curves.
3. All possible motions of a given mechanical system defining separate motions in their relation to other motions.

139. How did Functional Analysis come up ?

Consider the simple relation : $y = a x + b$. When $a = 2$ and $b = 3$, we get a particular case :

$$y = 2x + 3.$$

From the point of view of algebra, x and y are unknown numbers, but x and y can also be thought of as variables.

On the basis of the second idea, Des cartes produced his well known union of algebra and geometry, of an equation and a curve, which is one of the most important elements in the rise of analysis.

Likewise, the union of the concept of a varibable function with the ideas of contemporary algebra and geometry produced Functional Analysis.

140. What has been the impact of Functional Analysis ?

Just as analysis was necessary for the development of classical mechanics, so Functional Analysis provided new methods for the solution of the problems of mathematical physics and made available mathematical apparatus for the quantum mechanics of the atom.

141. What is the ultimate objective of Functional Analysis ?

Generality and unification are the two objectives Functional Analysis seeks to achieve.

These have been the distinctive features of the 20th century mathematics also.

142. What is the theory of Approximation of Functions ?

It is a theory which studies questions of the best approximate representation of general functions by various "simple" functions, especially by polynomials.

Polynomials are functions of the form

$$a_0 x^n + a_1 x^{n-1} + ... + a_{n-1} x + a_n ,$$

where $a_0, a_1 ,............, a_{n-1} , a_n$ are constants.

143. Where does the need for such approximation arise ?

It turns out that physical problems when expressed into mathematical parlance often show up complicated functions. These functions are not amenable to the process of analysis, especially differentiation and integration.

The theory lays down general foundations for the practical calculation of simple functions by which complicated functions can be approximately replaced without affecting the results significantly.

144. What is Tensor Analysis ?

Tensor Analysis is a branch of mathematics which deals with tensors.

145. What is a tensor ?

A tensor is a further generalisation of vectors, and is equal to zero, if and only if each of its conponents is equal to zero.

It has components in several different coordinate systems, the components in one system being related to those in the other systems by a definite law.

146. What is the aim of this study ?

Its chief aim is the investigation of relations that remain valid irrespective of the coodinate system in which they are expressed.

147. What are its applications ?

It was found to be particularly well suited to the needs of Einstein's theory of relativity, and has applications in many other branches of physics.

148. What is meant by discrete in Discrete Mathematics ?

Any set is said to be discrete if its members can be completely counted off as 1, 2, 3,.......

A discrete set need not necessarily consists of a finite number of elements.

The integers 1, 2, 3,....... form a discrete infinite set. All the particles of sand on the sea shore, all the atoms in the universe or all the stars are examples of discrete sets.

149. Which branches make up Discrete Mathematics ?

In Discrete Mathematics are included those branches of mathematics which focus mainly or entirely on discrete objects, including numerical analysis, abstract algebra, linear algebra, number thoery, graph theory and discrete probability.

150. What is meant by continuous as different from the discrete ?

The intuitive idea of continuous is that of an entity having no breaks.

According to our perception, motion is continous. A bullet moves evenly without breaks in its motion. A familiar example of continuity is that of all the points on a straight line.

Mathematical analysis is just based on the mathematics of continuity.

151. What are the implications of continuity in analysis ?

Continuity implies two assumptions :

'Infinite divisibility' and 'no-nextness'.

By 'infinite divisibility' is meant that whereas a line segment say one centimetre in length can be divided into a billion equal parts, a thin wire of one centimetre length cannot possibly be so divided. Probably it will turn into powder.

By 'no-nextness' is meant that whereas a point one centimetre away from one end of a line segment can be specified, the point just next to it either on the right or on the left cannot be so determined by any means whatsoever.

152. How do these assumptions affect mathematical results ?

Notwithstanding these assumptions, mathematical analysis, when applied to the physical world, yields fairly consistent results and makes fairly accurate predictions.

It is because the idealized representaiton serves as a sufficiently accurate description of the observable.

For example, the Sun and the Earth are, for gravitation, idealized as mere points and their size not taken into account.

153. How is it that the mathematics of continuity is far more developed than the mathematics of the discrete ?

'Space' and 'time' enter into most physical investigations, where they are assumed continuous. The reason why mathematics of the continuity is far more developed, therefore, seems to lie in their assumed continuity without which classical theories of physics since Newton would not have been possible.

Mathematics of the discrete, on the other hand, has no use for such assumption.

But the continuity of both space and time is a pure assumption.

154. In algebra, the concept of continuity is either completely absent or has a subordinate role. Is it not an intrinsic deficiency of algebra ?

No, on the contrary it is its significant characteristic.

In the real world the continuous and the discrete are found in inescapable unity of the opposites, for example, we think of water in a stream as flowing continuously though it is made up of discrete molecules. To know reality, it is sometimes necessary to dissect an object or idea into parts and to study these parts separately.

Algebra studies the discrete aspect of the objects of study.

155. What is the role of discrete mathematics in computer-related mathematics ?

The mathematics most important to the computer scientists belongs to the areas of discrete mathematics and not to calculus.

Contemporary research has therefore an increasing trend in the direction of discrete mathematics.

156. What is meant by Pure and Applied Mathematics ?

Mathematics can be subdivided into pure mathematics and applied mathematics.

Pure mathematics deals with the study of the abstract properties of mathematical entities and systems without regard to application.

Applied mathematics deals with solutions to practical problems in areas such as physics, economics, navigation and astronomy.

157. What is the criterion for this type of division ?

The division is in no way a rigid one. When a new application is found, a topic that was originally classified as pure mathematics may then become a part of applied mathematics.

158. What constitutes applied mathematics ?

Applied mathematics includes mechanics, theory of electricity and magnetism, relativity, theory of potential, thermodynamics and biomathematics.

Computer science, Probability, Statistics and Operations Research are often considered part of applied mathematics.

The term applied mathematics refers to the use of mathematical principles as tools in the fields of chemistry, biology, physics, engineering and social sciences.

159. What constitutes pure mathematics ?

The study and development of the principles of mathematics for their own sake rather than for their immediate use in other fields of science or knowledge is termed pure mathematics.

Abstract algebra, number theory and topology, are usually considered part of pure mathematics.

Often the study of problems in applied mathematics leads to new developments in pure mathematics, and theories developed as pure mathematics often find applications later.

No sharp line can be drawn between applied and pure mathematics.

160. How are pure and applied mathematics interdependent ?

The relation between pure and applied mathematics is such that neither can survive without the other.

The applied needs the pure for its tools and concepts but the pure also needs the applied for revitalisation. It is the applied that provides fresh challenges on which depends further growth and vitality of the pure.

161. What is included in the classical analysis ?

Classical analysis usually includes the following :

- Differential and Integral calculus,
- Differential equations,
- Differential geometry,
- Theory of functions of a complex variable,
- Vector and Tensor analysis,
- Calculus of Variations,
- Infinite series,
- Fourier series.

162. What constitutes modern analysis ?

Roughly speaking the following branches constitute modern analysis :

- Theory of functions of a real variable,
- Qualitative theory of differential equations,
- Theory of integral equations,
- Theory of Approximation of functions,
- Functional analysis.

Index

(The numbers refer to pages)